Where On Earth?

A Refreshing View of Geography

Donnat V. Grillet

A FIRESIDE BOOK
Published by Simon & Schuster
New York London Toronto Sydney Tokyo Singapore

F

FIRESIDE
Simon & Schuster Building
Rockefeller Center
1230 Avenue of the Americas
New York, New York 10020

Manufactured in the United States of America

3 5 7 9 10 8 6 4   Pbk.

Library of Congress Cataloging in Publication Data

Grillet, Donnat V.
Where on earth? / by Donnat V. Grillet.
p.   cm.
1. Geography—Popular works. I. Title.
G116.G73   1990
910—dc20   89-27474
CIP

0-671-76572-8

CONTENTS

DEDICATIONS

This book is dedicated to the winners of the
first two National Geography Bees
sponsored by the
National Geographic Society:

Jack Staddon, of Great Bend, Kansas, winner of
the 1989 National Geography Bee

Susannah Batko-Yovino, of Altoona, Pennsylvania,
winner of the 1990 National Geography Bee

Jack and Susannah really know the
answers to *Where On Earth?*

Presumably you are here . . .

. . . or here.

I Love Geography

We live on a glowing blue-green oasis hanging in the blackness of space, a blue marble revolving around the Sun at 67,000 miles per hour. Spaceship Earth.

And everywhere, absolutely everywhere, even on the bottom of the oceans, is geography.

Geography is in the very air we breathe. It connects us with everyone on the planet. It is the glue that binds together all impressions of where on Earth everything is.

Many people say that geography is their favorite subject. Yet about half of the American young people polled couldn't locate the United States on a map. Many said that Washington, D.C., is in the state of Washington or *is* the state of Washington. More than a quarter of young Texans could not name the country that lies "south of the border, down Mexico way."

It turns out that too many of us, younger and older, do not know where we are, much less who our neighbors are. There is a "geography gap" in our intelligence.

And so, as Gilbert M. Grosvenor, president of the National Geographic Society, has declared repeatedly, "Without geography, we're nowhere."

All of us *must* know where things are, where things happen, and, when possible, *why* things happen—geographically.

Earth is indeed an ever-changing world that seems "to lie before us like a land of dreams, So various, so beautiful, so new . . ." Which is why I've composed *Where On Earth?* It is designed to be highly informative, even entertaining.

Before you plunge into the multiple-choice questions and the book's many reader-involving features, allow me to share with you some thoughts I've had in my 25 years as a developer of geographic curricula and teaching methods.

It is a *serious* problem when we fail to achieve a minimum standard of competency in geography. It is a *critical* problem when national policy that relies on imprecise information is formulated.

Our history is dotted with examples reflecting general geographic illiteracy. The Dust Bowl of the 1930s reflected a general ignorance of topography and climate. The killer air-pollution attack on Donora, Pennsylvania, in 1948 reflected a lack of environmental knowledge. The discharge of liquid waste in a disposal well near Denver, Colorado, between 1962 and 1965 led to earthquakes, further demonstrating our lack of understanding of geology. Building structures on unstable ground contributed to devastation in the 1964 quake in Alaska.

Other recent geographically influenced events included the dust storm in the Iranian desert that aborted arguably the most important US military mission since Vietnam . . . the deaths near Mount Saint Helens's eruptions . . . the numerous landslides and mudslides and floods in California . . . the ongoing fight to clean up toxic waste and air pollution.

Formal education moved away from specialization after the Second World War. Science and math were pushed as America sped toward outer space. Geography became buried in the Social-Studies curriculum.

Three events in the 1960s planted the seeds for the resurrection of geography as a school subject nationally. First, the Peace Corps was established. Relatives, friends, and neighbors returned from distant places and told us about and showed us the outside world. Second, the horrors of Vietnam were on our TV sets every day, every night. Cambodia, Laos, and Vietnam were graphically displayed as casualty maps and troop-movement charts. The nightly news became, and is today, an extended geography lesson. Third, Americans walked on the Moon. That incredible sight was the ultimate geography lesson. We saw an Earthrise. We realized how isolated we are in space.

We have begun to think about the place we live. We don't like the purplish clouds of pollution over our cities and states and countries. We celebrated the first Earth Day. Every day lessons and the commentary of environmentalists grew in significance. Jacques Cousteau became a household name. The exploding awareness of the environment, this fragile interconnected web of life, has become a tremendous social, political, and cultural force.

Generations to come need the knowledge and the understanding to manage Earth's natural resources wisely. Every day, we use geographic knowledge to make important decisions about our personal well-being: where we should live, where we can build safe, secure, and desirable housing. Cultural facts such as quality of schools and convenient transportation routes require knowledge that is geographical.

You can imagine my delight when a joint committee of the National Council for Geographic Education and the Association of American Geographers published a booklet informing educational decision-makers about the need to institute, update, and enrich geography programs in America's schools—in *all* America's schools.

That booklet stimulated my preparation of *Where On Earth?* It led me to the 500 multiple-choice questions, the 60 or so maps, the features such as "Oddish Facts" and "What in the World? " and the dictionary of geographical terms.

I think this book will stir your curiosity and spark your interest. If you are studying for the National Geography Bee, sponsored annually nationwide by the National Geographic Society, I am sure that the information throughout this book will be of *great* help.

Our world changes every second. Political and economic events in eastern Europe . . . more foreign products in the American marketplace . . . the interplay of tectonic forces . . . the wiping out of the world's tropical rain forests . . . upheavals in the Middle East . . . Perhaps more than ever we all *need* to know *Where On Earth?*

Your Score

Correct Answers

550–500	You *really* know your geography. Bravo
499–400	Excellent
399–300	Very good. You have more than a passing knowledge of the world
299–150	Not marvelous, but good. Keep working at it
149–below	Time to hit the geography texts again

1. The most abundant food in the world is
    a. plankton
    b. krill
    c. maize
    d. rice

2. The state of _____ produces more grapes and more eggs than any other state in the United States.
    a. New York
    b. Pennsylvania
    c. Wisconsin
    d. California

3. One Canadian province produces more maple syrup annually than is produced in the entire United States. The province is
    a. British Columbia
    b. Alberta
    c. Manitoba
    d. Quebec

4. The cereal grain that is the basic food for more than half of the world's population is
    a. rice
    b. wheat
    c. oats
    d. barley

5. More wheat is imported by _____ than by any other country in the world.
    a. the Soviet Union
    b. the People's Republic of China
    c. Ethiopia
    d. England

6. A natural food widely consumed in the United States is indigestible to most of the adults in the world:
   a. the tomato
   b. the potato
   c. the peanut
   d. milk

7. The number-one fruit harvest in the United States is the
   a. orange
   b. grape
   c. pineapple
   d. peach

8. Asia's most productive rice granary is
   a. Myanmar
   b. Vietnam
   c. Malaysia
   d. Laos

9. More potatoes are produced in _____ than in any other country of the world.
   a. the Soviet Union
   b. the People's Republic of China
   c. Poland
   d. Argentina

10. Sub-Saharan Africans are utterly dependent on two drought- and heat-resistant cereals that thrive on dry, withered soils. One of the cereals is millet. The other is its cousin
    a. sorghum
    b. wheat
    c. rye
    d. oats

13

11. More of the arable surface of the globe is given over to _____ than to any other crop.

    a. rice

    b. wheat

    c. soybeans

    d. millet

12. The world's leading producer of corn is

    a. the United States

    b. Brazil

    c. Peru

    d. Spain

13. Famine in China in 1943 caused _____ million deaths.

    a. 1

    b. 3

    c. 8

    d. at least 13

14. "The Great Hunger" refers specifically to a famine in

    a. Ruanda-Urundi

    b. Ireland

    c. Ethiopia

    d. China

15. _____ is an indispensable food for the people of Japan, China, India, Indonesia, and Bangladesh.

    a. Wheat

    b. The soybean

    c. Rice

    d. Sorghum

16. About _____ plants have been identified as having edible parts.
    a. 50
    b. 85
    c. 440
    d. 6,000

17. The most densely populated agricultural country in the world is
    a. Belgium
    b. the Netherlands
    c. Bangladesh
    d. Japan

18. In the United States, the farm animal whose distribution is most closely related to the growing of corn is the
    a. horse
    b. swine
    c. cow
    d. chicken

19. The plant that is the major source of vegetable protein in the Japanese diet is
    a. rice
    b. the soybean
    c. bamboo
    d. the black bean

20. The fish-rich Grand Banks are located off the coast of
    a. Canada
    b. Florida
    c. Oregon
    d. Peru

21. During Europe's Industrial Revolution, this root crop from the western hemisphere played a key role in feeding the continent: the
    a. radish
    b. potato
    c. cabbage
    d. beet

22. Half of the agricultural output of the western African nation of Senegal is
    a. corn
    b. peanuts
    c. cotton
    d. grapes

23. From the 1500s to the 1800s, the cultivation of _____ created the prodigious demand in the Caribbean for African slaves.
    a. tobacco
    b. potatoes
    c. sugar cane
    d. dates

24. The plantation system is the agricultural arrangement used to produce much of the world's sugar cane and
    a. pineapples
    b. grapes
    c. oranges
    d. cotton

25. The tillable surface of the planet is _____ percent.
    a. 6
    b. 14
    c. 17
    d. 21

26. The philosopher _____ once said, "If you don't know the five grains, you don't know anything."

   a. Plato

   b. John Dewey

   c. Socrates

   d. Confucius

27. Seventy percent of the surface of the Earth is ocean. Seafood accounts for _____ percent of humankind's diet.

   a. 1

   b. 13

   c. 42

   d. almost 70

28. The _____ was termed "the miracle bean" by the ancient Chinese, and it has been rated the most remarkable of all legumes.

   a. soybean

   b. black bean

   c. red bean

   d. brown bean

29. The United States is the "breadbasket of the world." About _____ million acres are planted to wheat.

   a. 5

   b. 11

   c. 20

   d. 258

30. Corn's internationally preferred designation is

   a. maize

   b. sudd

   c. millet

   d. puszta

1. **(b)** Krill is the most abundant food in the world. Shrimplike, it is a principal diet for whales in the stormy waters around Antarctica.

2. **(d)** California produces both more grapes and more eggs than any other state in the United States.

3. **(d)** The Canadian province of Quebec produces more maple syrup annually than is produced in the entire United States.

4. **(a)** Rice is the cereal grain that is the basic food for more than half of the world's population. China, India, and Indonesia produce the most rice. Cereal grains supply about half the calories that people worldwide consume.

5. **(a)** The Soviet Union continues to be the world's number-one importer of wheat. The People's Republic of China is number two.

6. **(d)** Milk is indigestible to most of the world's adults.

7. **(d)** The orange is the largest fruit harvest in the United States.

8. **(a)** Myanmar, once called Burma, is Asia's most productive rice granary, but it has become an importer of rice as well.

9. **(a)** The Soviet Union produces about 25 percent of the world's potatoes. China produces 19 percent; Poland, 12 percent; the United States, 5 percent. The potato, which seems to have originated in Peru at least 8,000 years ago, was described by Noah Webster as one of the "greatest blessings bestowed on man by the Creator."

10. **(a)** Sorghum and millet are the cereals that hundreds of millions of sub-Saharan Africans depend on for living.

11. **(b)** Fields of wheat, which has been described as the quintessential foodstuff for more than a billion people, occupy about 600 million acres of the globe's arable surface, and so more of the arable surface of the globe is given over to wheat than to any other crop.

12. **(a)** The United States produces about 40 percent of the world's corn. The People's Republic of China and Brazil rank second and third in the production of corn.

13. **(b)** Three million Chinese perished in the famine of 1943. Between 100 BC and the early part of this century, there were at least 1,800 famines in China.

14. **(b)** The Irish famine of the late 1840s, caused by a potato blight, was the worst European disaster since the Black Death plague of five centuries earlier. About 1.5 million Irish died in the Great Hunger, and another 1.25 million emigrated, most of them to Great Britain or the United States, some to Australia.

15. **(c)** The people of Japan, China, India, Indonesia, and Bangladesh get 70 to 80 percent of all their calories from rice. India and China produce more than half of the world's rice and they consume more than 99 percent of what they grow.

16. **(d)** About 6,000 plants have parts that can be eaten. Of the billions of tons of food produced annually, more than 75 percent is of plant origin. It has been said that "we live in this world as guests of green plants."

17. **(c)** More than 100 million people are concentrated in Bangladesh, which is the size of the American state of Wisconsin, making the "Bengal nation" the most densely populated agricultural country in the world. (Per-capita income is $151 annually.)

18. **(b)** Of all farm animals, it is the distribution of the swine that is most closely related to the growing of corn in the United States.

19. **(b)** The soybean, or soya, is the plant that is the major source of vegetable protein in the Japanese diet.

20. **(a)** The Grand Banks, a spectacular fishing area, are located in the Atlantic Ocean off the Canadian island-province of Newfoundland.

21. **(b)** The potato played a key role in feeding Europe during the Industrial Revolution.

22. **(b)** Half of Senegal's agricultural output is peanuts.

23. **(c)** The cultivation of the imported plant sugar cane generated the great demand for African slaves in the Caribbean from the 1500s to the 1800s.

24. **(a)** The agricultural arrangement that has been used to produce much of the world's sugar cane and pineapples is the plantation system.

25. **(a)** The surface of the planet is 70 percent ocean and 24 percent untillable mountain, desert, and tundra. Only 6 percent is therefore tillable.

26. **(d)** The 5 sacred grains of Chinese antiquity to which Confucius (551–478 BC) alluded were rice, millet, sorghum, soybeans, and wheat.

27. **(a)** A mere 1 percent of the globe's diet is seafood. Humans have always relied on the produce of the soil for our principal sustenance.

28. **(a)** The "miracle" soybean bulges with protein—38 percent of its edible weight—and it has three times the protein of wheat, corn, or other cereals. It grows best in Manchuria.

29. **(c)** The United States has about 20 million acres planted to wheat; nearly three quarters is winter wheat. Kansas derives more revenue from the production of wheat than does any other state.

30. **(a)** It has been written that the history of the development of "maize," the internationally preferred designation of corn, "is inseparable from the history of the origin and development of civilization on the American continent." Corn was the most important grain crop that Native Americans taught the Pilgrims how to grow.

1. There are no penguin colonies in
    a. Africa
    b. Australia
    c. South America
    d. North America

2. About 80 percent of the world's reindeer are in
    a. Lapland
    b. Siberia
    c. Canada
    d. Alaska

3. The world's largest bird is the
    a. kiwi
    b. California condor
    c. ostrich
    d. albatross

4. In New Zealand, _____ outnumber people by more than 20 to one.
    a. kiwis
    b. platypuses
    c. sheep
    d. polar bears

5. There are more species of plants and animals in _____ than in any other place on the planet.
    a. Africa
    b. the Amazon rain forest
    c. Central America
    d. Micronesia

6. Wisents, the European bison, live in a protected forest in
   a. Kenya
   b. Poland
   c. Lapland
   d. the Alps

7. The giant tortoise, almost extinct, makes its home on the Tortoise Islands, which are more popularly known as the _____ Islands.
   a. Society
   b. Mariana
   c. Galápagos
   d. Cocos

8. The Guernsey breed of cattle originated on an island in
   a. the English Channel
   b. the Aegean Sea
   c. Delaware
   d. Hungary

9. Rattlesnakes are found only in North America, and there are _____ kinds.
   a. 7
   b. 9
   c. 29
   d. nearly 100

10. Native ants are found
    a. on all continents
    b. on all islands
    c. practically everywhere
    d. only in rain forests and North America

1. **(d)** There are penguin colonies only in Africa, Australia, South America, and Antarctica.

2. **(b)** Two million reindeer, or nearly 80 percent of the world's total, are in Siberia.

3. **(c)** The male ostrich, the world's largest bird, stands as tall as 8 feet and weighs up to 300 pounds.

4. **(c)** Sheep outnumber New Zealanders by more than 20 to one. The 2-island nation is the world's leading exporter of lamb.

5. **(b)** The Amazon rain forest contains more species of plants and animals than does any other place on Earth.

6. **(b)** The wisents, which stand 6 feet high at the shoulder when fully grown, live in a protected forest in Poland; they are nearly extinct.

7. **(c)** *Tortoise* in Spanish is *galápago*. The 3,075-square-mile Galápagos Islands, a province of the South American country of Ecuador, are on the Equator in the Pacific Ocean about 600 miles west of the mainland.

8. **(a)** The Guernsey breed of cattle originated on Guernsey, a 24-square-mile island in the English Channel.

9. **(c)** There are 29 kinds of rattlesnakes; the longest, the eastern diamondback, is 7 feet.

10. **(c)** Ants are found practically everywhere. Only Antarctica, Iceland, Greenland, and the part of Polynesia lying east of Tonga lack native ants.

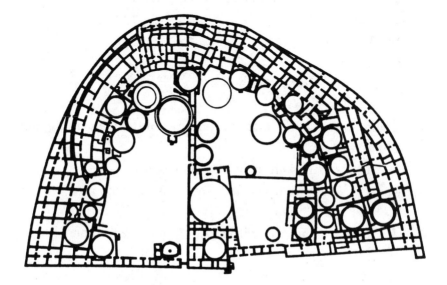

1. The world's oldest continuously inhabited city is
   a. Bombay
   b. Addis Ababa
   c. Damascus
   d. Baghdad

2. Paris, the capital of France, is the largest French-speaking city in the world. The second largest French-speaking city is
   a. Lyons
   b. Marseille
   c. Montreal
   d. Cayenne, French Guiana

3. Urumqui, one of the most exotic cities in the People's Republic of China, is
   a. the eastern terminus of the Lido Road
   b. located farther from the oceans than any other city in the world
   c. the highest city in Asia
   d. the first Asian city with a population of 2 million

4. "The Venice of the North" is the sobriquet of this European capital:
   a. Oslo, Norway
   b. Stockholm, Sweden
   c. Copenhagen, Denmark
   d. Helsinki, Finland

5. _____ is the highest capital city in Europe.
   a. Prague, Czechoslovakia,
   b. Geneva, Switzerland,
   c. Madrid, Spain,
   d. Copenhagen, Denmark,

6. The center of the capital of Mexico, Mexico City, is located on
    a. a built-up lake bed
    b. abandoned asbestos mines
    c. the ruins of 4 Aztec cities
    d. the graves of thousands of Spanish explorers

7. The world's largest Polynesian city is
    a. Auckland, New Zealand
    b. Honolulu, Hawaii
    c. Manila, the Philippines
    d. Djakarta, Java

8. The last town taken by the United States armed forces on the morning of Armistice Day in the First World War was
    a. Hamburg, Germany
    b. Kiel, Germany
    c. Berlin, Germany
    d. Stenay, France

9. Only two capitals are north of 60 degrees north latitude. One is Reykjavik, Iceland. The other is
    a. Ottawa, Canada
    b. Helsinki, Finland
    c. Moscow, USSR
    d. Oslo, Norway

10. The world's first "peopleless city" is
    a. Minsk
    b. Irkutsk
    c. Pinsk
    d. Chernobyl

11. The city of _____ has both the largest library in the world and the largest printing establishment in the world.
    a. Alexandria, Egypt,
    b. London, England,
    c. Beijing, the People's Republic of China,
    d. Washington, D.C.,

12. La Paz, Bolivia, which is 2.5 miles above sea level, is the highest capital city in the western hemisphere. The second highest is
    a. Quito, Ecuador
    b. Brasilia, Brazil
    c. Mexico City, Mexico
    d. Ottawa, Canada

13. There are _____ cities in Asia that have over a million people in each.
    a. 8
    b. 18
    c. 19
    d. 78

14. The Middle Eastern city that is the Holy City of 3 major religions is
    a. Istanbul
    b. Damascus
    c. Nazareth
    d. Jerusalem

15. The largest city in the West Indies is
    a. Kingston, Jamaica
    b. Port-au-Prince, Haiti
    c. Santo Domingo, Dominican Republic
    d. Havana, Cuba

16. The most populous city in the People's Republic of China is
    a. Xian
    b. Harbin
    c. Canton
    d. Shanghai

17. On the eve of the Second World War, Britain's Winston Churchill
    described this city as the "greatest target in the world . . . a fat cow
    tied up to attract the beasts of prey":
    a. London
    b. New York
    c. Paris
    d. Berlin

18. The west African town of Tombouctou, in Mali, was called Timbuktu
    when it was a center of Muslim culture and reached its height of
    prosperity as a commercial and cultural center under Songhai rule.
    Its population was once about a million. Today, Tombouctou's
    population is about
    a. 20,000
    b. 200,000
    c. 2,000,000
    d. 3,250,000

19. The Soviet city of Stalingrad, named for the dictator Joseph Stalin,
    marked the farthest eastern advance of Hitler's army during the
    Second World War. Stalingrad's name has been changed to
    a. Volgograd
    b. Leningrad
    c. Gorbachevgrad
    d. Minsk

20. The capital of Bangladesh is

   a. Chittagong

   b. Dacca

   c. Bogra

   d. Rangpur

21. The "archeological capital of South America" is

   a. Lima

   b. Valparaiso

   c. Montevideo

   d. Cuzco

22. The strategically important seaport of Dakar on the western tip of Africa is the capital of

   a. Senegal

   b. the Ivory Coast

   c. Sierra Leone

   d. Mauritania

23. The first cities in the Americas were in

   a. eastern Canada

   b. Mexico

   c. the southwest United States

   d. Peru

24. At mid-twentieth century, there were only 7 urban centers each with more than 5 million people: New York, London, Paris, Germany's Rhein-Ruhr complex, Tokyo-Yokohama, Shanghai, and Buenos Aires. Today, at least _____ urban centers boast more than 5 million people each.

   a. 17   **c.** 29

   b. 22   **d.** 34

25. At last count, each of _____ cities in North America, the third
    largest continent, had more than a million residents.
    a. 8
    b. 13
    c. 39
    d. 83

26. The main industrial, commercial, and transportation center of
    central China is
    a. Wuhan
    b. Chungking
    c. Lanchou
    d. Chengchou

27. Of the following capital cities, the closest to Ankara, the capital of
    Turkey, is
    a. Moscow
    b. Teheran
    c. Tripoli
    d. Paris

28. Of these capital cities, the farthest from Madrid, the capital of
    Spain, is
    a. Rabat
    b. Berlin
    c. Cairo
    d. Helsinki

29. Vientiane is the present capital and Luang Prabang is the old royal
    capital of
    a. Thailand   c. Laos
    b. Cambodia   d. India

1. **(c)** The Syrian capital of Damascus, which was built in an oasis, is considered the world's oldest (at least about 4,000 years) continuously inhabited city, and it is the oldest capital city in the world.

2. **(c)** Montreal, Canada, is second only to Paris as the largest French-speaking city in the world.

3. **(b)** Urumqui, the capital of China's Xinjiang Uygur Autonomous Region, is located farther (2,500 miles) from the oceans than any other city in the world. It is situated in an oasis bounded by desert in the south and by snow-covered mountains in the north.

4. **(b)** The Swedish seaport city of Stockholm, "the Venice of the North," was built on a peninsula and several islands. Its population is around 1.4 million.

5. **(c)** Madrid, the capital of Spain, is the highest capital city (2,150 feet) in Europe.

6. **(a)** A built-up lake bed underlies the center of Mexico City.

7. **(a)** Auckland, New Zealand's largest city, is the world's largest Polynesian city, with over 90,000 Maoris, Samoans, and Cook Islanders.

8. **(d)** Stenay, on the Meuse River and 26 miles north-northwest of Verdun, was the last town recaptured by the US military on the morning that Germany surrendered and the First World War was over: November 11, 1918.

9. **(b)** Reykjavik and Helsinki are both north of 60 degrees north latitude, with the Icelandic capital a few degrees farther north than the Finnish capital.

10. **(d)** Because of the accident at its nuclear power plant, in April, 1986, Chernobyl in the Soviet Union was declared unsafe for habitation—it is the world's first "peopleless city."

11. **(d)** The Library of Congress is the largest library in the world, and the US Government Printing Office is the largest printing establishment in the world—both are located in Washington, D.C.

12. **(a)** The capital of Ecuador, Quito, is 9,249 feet above sea level, making it second only to La Paz, Bolivia, as the highest capital city in the western hemisphere.

13. **(d)** Asia, the largest continent in the world, has a population of more than 2.5 billion, 78 of its cities having more than a million people each.

14. **(d)** The Israeli capital, Jerusalem, is the Holy City of Jews, Christians, and Muslims.

15. **(d)** Havana, the seaport capital of Cuba, 90 miles south-southwest of Key West, Florida, is the largest city in the West Indies. It has a population of just over a million, and its harbor is one of the best in the hemisphere.

16. **(d)** Its population of 12 million makes Shanghai the most populous city in the People's Republic of China, the world's most populous nation (more than a billion people).

17. **(a)** Winston Churchill was convinced that Hitler's fast-growing air force would find Britain's capital city, London, "the greatest target in the world."

18. **(a)** The population of Tombouctou, in west Africa's Mali, is today a little more than 20,000. The city declined rapidly in the sixteenth, seventeenth, and eighteenth centuries.

19. **(a)** Volgograd, at the eastern terminus of the Volga-Don Canal, is a major river port and railroad junction in the Soviet Union. Its population is around a million. Volgograd was Stalingrad, named for the Soviet dictator Joseph Stalin, from 1925 to 1961, when his brutality against Soviet citizens was revealed, prompting the name-change.

20. **(d)** Dacca, the capital of Bangladesh, was once the capital of the Mogul province of East Bengal and of the former British province of Eastern Bengal and Assam.

21. **(d)** The city of Cuzco, 350 miles southeast of Lima, Peru, was once the capital of the Incan Empire and is today celebrated as the "archeological capital of South America." Founded in the eleventh century, Cuzco was the "City of the Sun."

22. **(a)** Dakar, with one of the best harbors on Africa's Atlantic coast, is the capital of Senegal.

23. **(b)** The first cities in the Americas arose in Mexico about 500 BC. About a thousand years later, the first planned city, Teotihuacan, also in Mexico, housed as many as 200,000 people. (Athens in the golden age of Greece had a population of about 150,000. The first high-density dwellings in the western world were in Rome, which around the year 100 AD had a population of about 1 million.)

24. **(d)** There are today in the world at least 34 urban centers each with more than 5 million people. The United Nations predicts that by the year 2025 there will be 93, about 80 of them in the emerging nations.

25. **(c)** Each of 39 cities in North America, from the Arctic to Panama and including Greenland, the world's largest island, reported in the last count a population of more than a million.

26. **(a)** Wuhan, in east-central China, is the region's main industrial, commercial, and transportation center. Its population is around 6 million.

27. **(b)** Teheran is closer to Ankara than are Moscow, Tripoli, or Paris.

28. **(d)** The Finnish capital of Helsinki is farther from Madrid than are Rabat, Cairo, or Berlin.

29. **(c)** Vientiane, noted for its canals and houses built on stilts, became the capital of the former southeast Asian French protectorate of Laos in 1899. Luang Prabang, 130 miles to the northwest, was Laos's royal capital; it was founded by Indian Buddhist missionaries. Both cities are on the Mekong River.

- The Amazon River, in South America, contains more water than the Nile, Yangtze, and Mississippi Rivers combined—nearly one fifth of all the freshwater that runs over the surface of the Earth.
- One in 5 American adults believes that the Earth goes around the Sun once a day (instead of once a year).
- The Malaspina Glacier, in the Saint Elias Mountains, which extend from Alaska into Canada, is larger than the state of Rhode Island.
- Reno, Nevada, lies about 89 miles farther west than Los Angeles, California, and 164 miles farther west than Tijuana, Mexico.
- The Pacific Ocean is larger than all the land in the world.
- All of the United States could be placed within Africa's Sahara desert.
- There are some 40 countries and 300 million people in black Africa, or sub-Saharan Africa, which excludes the northern part of the continent (considered part of the Arab world) and also the Republic of South Africa.
- Vladivostok is on the "wrong" side of the Pacific Ocean. The Soviet Union's Pacific port is at a latitude about midway between San Francisco and Seattle on the Pacific coast of the United States; but, unlike the 2 American ports, Vladivostok needs icebreakers to be kept open through the winter months.
- Most of Australia's mammals are marsupials.
- Modern civilization has been able to extend its life by making use of ancient, stored forms of energy. We're running out of time.
- Rain forests the size of New York City's Central Park are destroyed every 8 minutes.
- Africa by the year 2025 is expected to have a population greater than the combined populations of Europe, North America, and South America.
- Three species become extinct every day.
- Half of California's college students who were polled did not know where Japan was on the map.
- The world's largest penal colony is in New York City.
- There have been no private automobiles in Albania during the hardline Communistic regime.

- Since 1901, successive Australian governments have tended to adhere to a "white Australia"-only policy.
- The state of Washington is larger than all 6 New England states combined.
- There is virtually no coal in South America.
- Sea level is the "highest point" in the United Arab Emirates, a federation of 7 independent states on the Persian Gulf, each with its own emir, or Arab chieftain.
- Boston, Massachusetts, is nearer by sea to Rio de Janeiro, the former capital of Brazil, than is New Orleans, Louisiania, located on the Gulf of Mexico.
- Indonesia would be the next land you would reach if you traveled in a straight line, that is, on the great-circle route, from Miami, Florida, through Portland, Oregon, and continued on the same line out to sea.
- It is hard to start a fire in La Paz, Bolivia, the world's highest capital, because there is little oxygen there, and oxygen is needed for combustion.
- The animal life of the Galápagos Islands in the Pacific Ocean inspired much of Charles Darwin's theory of evolution.
- The Caspian Sea, in the Soviet Union and Iran, is the largest *lake* in the world.
- Denmark is a peninsula and 400 islands.
- The ostrich was hunted to near extinction for its decorative feathers.
- A shallow prehistoric sea once covered Arizona where the Grand Canyon, gouged out by the Colorado River, is now a popular landmark.
- Jamaica, Guyana, and Suriname all produce bauxite, the principal ore of aluminum.
- Seventeen mountain peaks soar more than 10,000 feet over New Zealand's South Island.
- Visitors to Antarctica call the continent "the Ice."
- The greatest cod-fishing region of the world, the Grand Banks, in the Atlantic Ocean east and south of Newfoundland, Canada, is made dangerous by fog and icebergs.

- Islands dot the vast expanse of the Pacific Ocean. Some are atolls, and some are the tips of volcanoes that push up through the water. Rings of coral surround lagoons, which remain where volcanic peaks have sunk back into the sea.
- During their exploration of the Louisiana Territory and the unknown northwest, Meriwether Lewis and William Clark discovered 24 Native American nations (with which they dealt with an unmatched record of decency), 178 plants, and 122 animals previously unknown to Americans in the eastern United States. They set the stage for the new nation's westward expansion.
- Three types of continental climate—warm summer, cool summer, and subarctic—occur only in the northern hemisphere.
- There are more Spanish-speaking people in Mexico than there are in Spain.
- The Appalachian Mountains, in the eastern United States, were formed between 280 million and 300 million years ago. The Rocky Mountains, in the western part of the country, were formed "only" 60 million to 80 million years ago.
- Roughly one half of the Netherlands is below sea level.
- The only mammals that fly are bats.
- The American buffalo is actually the bison. Real, true buffalo live only in Africa and in Asia.
- The world's longest underwater mountain range is the Mid-Atlantic Ridge. It rises in the center of the Atlantic Ocean and extends from Iceland almost to Antarctica.
- Yes, snow falls in the tropics. It is common at high elevations in tropical regions all around the world.
- Antarctica is technically a desert (although a very cold one) because it has little or no precipitation.
- The Greeks of antiquity believed that any person going to the Equator would turn black.
- There are no large lakes in Spain.

- Virtually the whole continent of South America is east of Savannah, Georgia.
- Places with a tropical wet climate, such as equatorial west Africa and Hawaii, have the most predictable weather on the planet: It rains nearly every afternoon.
- The continent of Asia covers more area than the continents of North America, Europe, and Australia combined.
- Surface ocean currents in the midlatitudes of the northern hemisphere usually circulate clockwise, and in the midlatitudes of the southern hemisphere they usually circulate counterclockwise.
- The Caribbean end of the Panama Canal is west of the Pacific Ocean end.
- The Ural Mountains east of Moscow break the Great Northern European Plain, which stretches eastward from western France.
- Labor migration is a key factor in the decline of agriculture in the southwest Asian republic of Yemen, on the southern Arabian Peninsula. Some 700,000 Yemenis work in the neighboring country of Saudi Arabia.
- Weather on both sides of a continent becomes generally cooler as latitude increases.
- The time difference (3 hours, 50 minutes) between the most eastern and the most western parts of Alaska is greater than the time difference between New York on the east coast of the United States and San Francisco on the west coast.
- Asia, the world's largest continent by far, has—appropriately—some of the world's coldest and hottest climates, longest rivers, highest mountains, and largest deserts. Its far northern lands are snow-covered three quarters of the year, and its far southern lands include hot, steamy jungles.
- The planet has 3 layers: an outer crust, a middle rock mantle, and an inner core.

1. If there were no water in the 5 Great Lakes in North America, late
    autumns and winters along the area of their eastern shores would
    be colder
    a. but less snowy
    b. by at least 50 degrees
    c. and snowier
    d. for about a month, then much warmer, maybe by 50 degrees

2. The Holy City of Jerusalem, now the capital of Israel, at the eastern
    Mediterranean, once had a skein of no rain in the month of July for
    _____ years.
    a. 3
    b. 13
    c. 68
    d. more than 100

3. The 5 warmest years around the globe in the past 13 decades
    occurred in the same 10-year period:
    a. the 1980s
    b. the 1940s
    c. the 1930s
    d. the 1880s

4. Clouds were first classified into 3 main groups—cirrus, stratus, and
    cumulus—about _____ years ago.
    a. 50
    b. 100
    c. 200
    d. 2,750

5. The lowest temperature ever recorded on the face of the Earth was in
   a. Antarctica
   b. Siberia
   c. Alaska
   d. Norway

6. Violent storms originating in the Himalayas give _____ its nickname (in Dzongkha) "Land of the Thunder Dragon."
   a. Tibet
   b. Bhutan
   c. Nepal
   d. India

7. _____ experienced the highest wind velocity ever recorded in North America.
   a. Boulder, Colorado,
   b. Mount Washington, New Hampshire,
   c. Hudson Bay, Canada,
   d. Edmonton, Alberta (Canada),

8. On a typical day, there is an average temperature difference of _____ degrees Fahrenheit between the floor of the 280-mile-long Grand Canyon, in the northwest corner of Arizona, and its highest rim.
   a. 5
   b. 20
   c. 35
   d. 50

9. The southwest coast of the Hawaiian island of Kauai rarely receives more than 20 inches of rain a year. Twenty miles away, a mountaintop receives more than _____ inches of rain a year.
    a. 50
    b. 100
    c. 400
    d. 1,200

10. In winter, the most active seaport in Canada is
    a. Quebec
    b. Vancouver
    c. Halifax
    d. Prince Edward Island

11. The rainiest place in the 48 contiguous states of the United States is
    a. Olympic Peninsula, in Washington
    b. Great Dismal Swamp, in Virginia and North Carolina
    c. Alligator Swamp, in eastern North Carolina
    d. Everglades National Park

12. In the climate called Mediterranean, it rains most often in
    a. spring
    b. summer
    c. autumn
    d. winter

13. A warm summer day at Antarctica is about
    a. 20° F below zero
    b. 1° F above zero
    c. 32° F above zero
    d. 45° F above zero

14. The hottest place ever in North America and the driest place
    annually are both in the same state of the United States:
    a. Utah
    b. Nevada
    c. Arizona
    d. California

15. Earth's prevailing winds, like its rotation, are
    a. west to east
    b. east to west
    c. north to south
    d. south to north

16. It doesn't rain 9 months of the year in the harsh, dry, hot, vast
    Kalahari Desert, which is in
    a. Chile
    b. southern Africa
    c. Siberia
    d. Mongolia

17. The most powerful winds ever recorded in the Caribbean were
    _____ miles per hour.
    a. 73
    b. 93
    c. 117
    d. 170

18. Jet streams are long, narrow belts of rapidly moving air that occur
    at altitudes of 6 to _____ miles.
    a. 7
    b. 8
    c. 9
    d. 10

19. When the waters of the Atlantic Ocean near the Equator around the Cape Verde Islands, off the west African coast, are _____, they breed most of the 100 or so disturbances that the National Hurricane Center tracks every year.
    a. warm
    b. boiling hot
    c. cool
    d. ice-cold

20. The world's highest recorded temperature was in
    a. Java
    b. Brazil
    c. Libya
    d. the Caroline Islands

21. The hot wind that blows northward from the Sahara desert to the northern Mediterranean coasts is called the
    a. chinook
    b. sirocco
    c. xerophyte
    d. zenithal

22. The horse latitudes are
    a. subtropical belts of high atmospheric pressure over the oceans
    b. skies covered with cirrocumulus clouds
    c. lava flows
    d. cold waves, or busters, in western Australia

1. (a) Late autumns and winters along the area of the 5 Great Lakes, if they were bone-dry, would be colder but less snowy.

2. (d) For a recorded stretch of more than 100 years it didn't rain in Jerusalem in the month of July.

3. (a) All of the 5 warmest years in the past 130 years occurred around the globe in the 1980s.

4. (c) British biologist Luke Howard developed the system for naming clouds around the year 1800.

5. (a) The record reading of minus 126.9° F was recorded at Vostok, Antarctica.

6. (b) Violent storms originating in the Himalayas give Bhutan its name (in Dzongkha) "Land of the Thunder Dragon."

7. (b) The highest wind velocity ever recorded in North America was measured on Mount Washington, New Hampshire, on April 12, 1934: 231 miles per hour.

8. (d) The Grand Canyon, which has, on a typical day, a temperature difference of 50° F between its floor and its highest rim, is more than a mile deep in places.

9. (c) The southwest coast of Kauai, Hawaii, rarely receives more than 20 inches of rain a year because it is on the lee side of the island. It is in the rain shadow of a mountain only 20 miles away that receives more than 400 inches of rain a year.

10. (c) Halifax, in the central part of the southern coast of the province of Nova Scotia, is Canada's most active seaport in winter. It has been a Canadian naval base on the Atlantic Ocean since 1910.

11. (a) A peninsula in the western part of the state of Washington, Olympic is bounded on the west by the Pacific Ocean and is the rainiest place in the 48 contiguous states of the United States.

12. (d) It rains most often in winter in a Mediterranean climate.

13. (b) A warm summer day at Antarctica is about 1° F above zero. The average daily temperature of the six-month Antarctic winter is 70° F below zero.

14. **(d)** It was 134° F above zero in Death Valley, California, on July 10, 1913, and the annual rainfall in that eastern California region averages 1.5 inches a year.

15. **(a)** Earth's prevailing winds and rotation are west to east.

16. **(b)** The Kalahari Desert, in southern Africa, is home to about 30,000 Kung people.

17. **(d)** The 170-mile winds of hurricane Allen, August 1980, raged in the mightiest Caribbean storm ever recorded.

18. **(d)** The jet streams occur at altitudes of 6 to 10 miles.

19. **(a)** Warm ocean waters breed weather disturbances. Cool ocean waters are poor nourishment for tropical storms.

20. **(c)** The world's highest recorded temperature was 136.4° F in Azizia, Libya.

21. **(b)** The sirocco—or the simoom, the gharbi, or the leveche—is the name of the hot wind that blows northward from the Sahara to the northern Mediterranean coasts.

22. **(a)** The horse latitudes, which are situated between the trade winds and the westerlies in both the northern and southern hemispheres, are subtropical belts of high atmospheric pressure over the oceans. They are regions of calms and light variable winds.

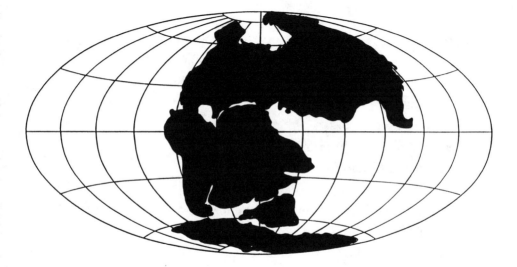

1. Asia, at 17,139,445 square miles, is the world's largest continent. The second largest continent is
   a. Africa
   b. South America
   c. Australia
   d. Antarctica

2. There are _____ independent countries in Africa.
   a. 13
   b. 43
   c. 51
   d. 70

3. Most of the first 170,000 settlers of Australia were
   a. convicts
   b. gold prospectors
   c. widows and children
   d. retired sailors from the British navy

4. Christopher Columbus was sailing for the Asian islands of Cipango, which was probably
   a. the Maldives
   b. Indonesia
   c. the Philippines
   d. Japan

5. There are _____ independent countries in Europe.
   a. 8
   b. 15
   c. 27
   d. 32

6. The only Portuguese-speaking nation in the Americas is
    a. Honduras
    b. Brazil
    c. El Salvador
    d. Chile

7. The continent with the lowest average annual amount of
    precipitation is
    a. Africa
    b. Australia
    c. Antarctica
    d. Europe

8. The geographic center of North America is in the state of
    a. Missouri
    b. Colorado
    c. Iowa
    d. North Dakota

9. There was once a supercontinent that combined India, Antarctica,
    Australia, South America, and
    a. Africa
    b. Central America
    c. the Philippines
    d. the United Kingdom

10. _____ countries of Europe are landlocked.
    a. Two
    b. Three
    c. Nine
    d. Fifteen

11. North America is located in these 2 hemispheres of the globe:
     a. northern
     b. northern and western
     c. southern
     d. northern and eastern

12. There are _____ independent nations in South America.
     a. 7
     b. 12
     c. 17
     d. 23

13. The most common tropical disease in Central America is
     a. cholera
     b. malaria
     c. chagas
     d. yellow fever

14. Only one country in South America has coastlines on both the Pacific Ocean and the Caribbean Sea:
     a. Colombia
     b. Venezuela
     c. Guyana
     d. Suriname

15. South America is the world's fourth largest continent; in population it ranks
     a. first
     b. second
     c. fifth
     d. last

16. The lowest point in Europe is
    a. the Caspian Sea
    b. the Po Valley
    c. the Black Forest
    d. the Seine Gut

17. The largest empire in history was
    a. Alexander the Great's
    b. the Mongols'
    c. Caesar's
    d. the Soviet Union's "evil" empire

18. Only 1 country in South America is completely outside the tropical latitudes:
    a. Uruguay
    b. Argentina
    c. Chile
    d. Paraguay

19. The opium-producing area of Laos, Myanmar, and Thailand, in southeast Asia, is known as
    a. the golden triangle
    b. the melting pot
    c. the Field of Dreams
    d. the snow-white circle

20. The division between Europe and Asia is marked by
    a. the Ural Mountains
    b. Poland's western boundary
    c. Poland's eastern boundary
    d. the northern bank of the Danube River

21. The British Empire in the East was begun in
    a. Afghanistan
    b. Siam
    c. Singapore
    d. west Bengal

22. Nearly half of the world's population lives on 1 continent:
    a. South America
    b. Africa
    c. Australia
    d. Asia

23. The only southeast Asian country never colonized by a European power is
    a. Singapore
    b. India
    c. Vietnam
    d. Thailand

1. **(a)** Africa, the world's second largest continent, is 11,677,239 square miles. It is the only continent in all 4 hemispheres: northern, southern, eastern, western.

2. **(c)** There are 51 independent countries in Africa.

3. **(a)** Australia was established as a penal colony by England, and most of the first 170,000 settlers were convicts.

4. **(d)** The islands of Cipango, described by Marco Polo with the spelling Zipangu, were east of Asia and today are generally identified with Japan.

5. **(d)** Europe has 32 independent countries plus portions of Turkey and the Soviet Union.

6. **(b)** Brazil, once a Portuguese colony, is the only Portuguese-speaking nation in the western continents.

7. **(c)** Antarctica is the continent with the lowest average amount of precipitation year in and year out.

8. **(d)** The exact geographic center of North America is the city of Rugby, about 65 miles east of Minot, in north-central North Dakota, the Flickertail State.

9. **(a)** Africa, South America, Australia, Antarctica, and India were all once part of the supercontinent Gondwanaland after the breakup of Pangaea.

10. **(c)** The 9 landlocked countries of Europe are Andorra, Austria, Czechoslovakia, Hungary, Liechtenstein, Luxembourg, San Marino, Switzerland, and Vatican City.

11. **(b)** North America is in the northern and western hemispheres of the globe.

12. **(b)** There are 12 independent nations in South America.

13. **(c)** Chagas, the most common tropical disease in Central America, is an incurable parasitic illness. Like AIDS, it damages the immune system, and kills usually after a long delay.

14. **(a)** Colombia is the only South American country with borders on both the Pacific Ocean and the Caribbean Sea.

15. **(c)** South America, with about 250 million people, is the fifth largest continent in population.

16. **(a)** The Caspian Sea, which is about 92 feet below sea level, is the lowest point in Europe.

17. **(b)** At its peak, the Mongol Empire stretched from the Pacific Ocean across most of continental Asia to the Danube River in Europe and to the Persian Gulf in the Middle East—it was history's largest empire.

18. **(a)** Northern Uruguay is about 30 degrees south of the Equator.

19. **(a)** About 20 percent of the heroin that reaches the United States originates as opium in "the golden triangle" of Laos, Myanmar, and Thailand, in southeast Asia.

20. **(a)** The Urals, a mountain chain in the middle of a thousand-mile flatland formed 225 million years ago, are east of Moscow and mark the division between Europe and Asia. The Urals have been so worn over the ages that they're no higher than 6,000 feet.

21. **(d)** Plassey, a village in west Bengal about 80 miles north of Calcutta in northeast India, was the site where the British Empire in the East began.

22. **(d)** Nearly half of the world's population lives in Asia.

23. **(d)** Once called Thailand, then Siam, the country reverted to Thailand in 1939, and it is Thailand, never colonized, today.

abyss. a deep fissure in the earth; a bottomless gulf; the ocean depths

Aeolian. relating to, or caused by, the wind

agonic line. an imaginary line on the Earth's surface along which true north and magnetic north are identical, and a compass needle makes no angle with the meridian

agriculture. the science and art of farming; the work of cultivating the soil, producing crops, and raising livestock

air mass. a huge, uniform body of air having the properties of its place of origin

alkalic. designating rocks that contain an unusually large amount of sodium and potassium minerals

alluvial fan. a gradually sloping mass of sand, clay, or the like that widens out like a fan from the place where a stream slows down little by little as it enters a plain

altitude. height, especially the height of a thing above the Earth's surface or above sea level

aquifer. an underground layer of porous rock, sand, etc., containing water, into which wells can be sunk

archipelago. a sea with many islands; a group or chain of many islands

atlas. a book of maps

atmosphere. the gaseous envelope (air) surrounding the Earth to a height of about 621 miles

atoll. a ring-shaped coral island nearly or completely surrounding a lagoon

avalanche. a mass of loosened snow, earth, rocks, etc., suddenly and swiftly sliding down a mountain, often growing as it descends

axis. a real or imaginary straight line on which an object rotates or is regarded as rotating, such as the axis of a planet

barometer. an instrument for measuring atmospheric pressure. It is used in forecasting changes in the weather or finding height above sea level

basin. a depression in the surface of the planet; all the land drained by a river and its branches

bay. a part of a sea or lake, indenting the shoreline; a wide inlet not so large as a gulf; any level land area making an indentation, as into a woods or a range of hills

bayou. in some parts of the southern United States, a sluggish, marshy inlet or outlet of a lake, river, etc.

biome. any of several major life zones of interrelated plants and animals determined by the climate, as deciduous forest or desert

biosphere. the zone of planet Earth where life naturally occurs, extending from the deep crust to the lower atmosphere; the living organisms of the Earth

bluff. a high, steep, broad-faced bank or cliff

bog. a wet, spongy ground, characterized by decaying mosses that form peat; a small marsh or swamp

boundary. any line or thing marking a limit

butte. a steep hill standing alone in a plain, especially in the western United States; a small mesa

caldera. a broad, craterlike basin of a volcano, formed by an explosion or by collapse of the cone

canal. an artificial waterway for transportation or irrigation

canyon. a long, narrow valley between high cliffs, often with a stream flowing through it

cape. a piece of land projecting into a body of water

cartography. the art or work of making maps or charts

cave. a hollow place inside the Earth, usually an opening, as in a hillside, extending back horizontally; a cavern

channel. the bed of a running stream or river; a body of water joining two larger bodies of water

city. a center of population larger or more important than a town or village

cliff. a high, steep face of rock, especially one on a coast; a precipice

climate. the prevailing or average weather conditions of a place, as determined by the temperature and meteorological changes over a period of years

cloud. a visible mass of tiny, condensed water droplets or ice crystals suspended in the atmosphere

coast. land alongside the sea; a seashore

compass. a device that indicates direction, one of the most important instruments in navigation

cone. the peak of a volcano

continent. any of the 7 main large land areas of Earth

Continental Divide. ridge of the Rocky Mountains forming a North American watershed that separates rivers flowing in an easterly direction from those flowing in a westerly direction

continental shelf. the submerged shelf of land that slopes gradually from the exposed edge of a continent for a variable distance to the point where the steeper descent (the continental slope) to the ocean bottom begins, commonly at a depth of about 600 feet

coral reef. a reef, or ridge, in relatively shallow, tropical seas composed chiefly of the skeletons of coral, which is the hard, stony secretion of certain marine polyps

core. one of the planet's 3 main layers, which are core, mantle, and crust. The solid, inner part of the core is thought to be made up of superhot, iron-rich material. The outer part of the core is probably molten iron and nickel. The entire core is about 4,350 miles in diameter, more than half the diameter of the planet

country. an area of land; a region; the whole land or territory of a nation or state; the people of a nation or state

crater. a bowl-shaped cavity, as at the mouth of a volcano

crevasse. a deep crack or fissure, especially in a glacier

crust. the solid, rocky, outer portion or shell of the planet; the lithosphere

current. a flow of water or air, especially when strong or swift, in a definite direction; specifically, such a flow within a larger body of water or mass of air

cyclone. loosely, a windstorm with a violent, whirling movement; a tornado or hurricane

date line. a boundary running from Pole to Pole. Roughly following the 180th meridian, which is halfway around the globe from the Prime Meridian (0°), it is the beginning of the calendar day

delta. a deposit of sand and soil, often roughly triangular, formed at the mouth of some rivers, as of the Nile

desert. an uncultivated region without inhabitants; a wilderness; a dry, barren, sandy region, naturally incapable of supporting almost any life

dike. an embankment or dam made to prevent flooding by the sea or by a river

drought. a prolonged period of dry weather; lack of rain

dune. a rounded hill or ridge of sand heaped up by the action of the wind

earthquake. a shaking or trembling of the crust of the earth, caused by underground volcanic forces or by the breaking and shifting of rock beneath the surface

ecology. the study of the relationship and adjustment of humans or other organisms to their geographical and social environment

ecosystem. a system made up of a community of animals, plants, and bacteria and its interrelated physical and chemical environment

elevation. height above a surface, as of the Earth

emigration. the act of emigrating, leaving one country or region to settle in another

energy. those resources, such as petroleum, coal, gas, wind, nuclear fuel, and sunlight, from which energy in the form of electricity, heat, etc., can be produced

environment. all the conditions, circumstances, and influences surrounding, and affecting the development of, an organism or group of organisms

Equator. an imaginary circle around the planet, equally distant at all points from both the North Pole and the South Pole. It divides the Earth's surface into the northern and southern hemispheres

equinox. the time when the Sun crosses the Equator, making night and day of equal length in all parts of the Earth

erode. to eat into; to wear away; to form by wearing away gradually

eruption. a bursting forth or out, as of lava from a volcano

escarpment. a steep slope or cliff formed by erosion or, less often, by faulting

estuary. an inlet or arm of the sea; especially, the lower portion or wide mouth of a river, where the salty tide meets the freshwater current

evapotranspiration. the total water loss from the soil, including that by direct evaporation and that by transpiration from the surfaces of plants

fall line. the topographical line indicating the beginning of a plateau, usually marked by many waterfalls and rapids

famine. an acute and general shortage of food, or a period of such shortage

fault. a fracture or zone of fractures in rock strata, together with movement that displaces the sides relative to one another

fjord. a narrow inlet or arm of the sea bordered by steep cliffs

flood. an overflowing of water on an area normally dry; the flowing in of water from the sea as the tide rises

fog. a large mass of water vapor condensed to fine particles, at or just above the Earth's surface; thick, obscuring mist

forest. a thick growth of trees and underbrush covering an extensive tract of land

fossil. any hardened remains or imprints of plant or animal life of some previous geologic period, preserved in the Earth's crust, such as petrified wood and dinosaur footprints

fossil fuels. fuels such as coal, petroleum, and natural gas, found underground in deposits formed in a previous geologic period

front. the boundary between 2 air masses of different temperature and humidity

gap. a mountain pass, cleft, or ravine

geology. the science dealing with the physical nature and history of the Earth, including the structure and development of its crust, the composition of its interior, individual rock types, etc.

geotectonic. having to do with the structure, distribution, shape, etc., of rock bodies, and with the structural disturbances and alterations of the Earth's crust that produced them

geyser. a spring from which columns of boiling water and steam gush into the air at intervals

glacier. a large mass of ice and snow that forms in areas where the rate of snowfall constantly exceeds the rate at which the snow melts. It moves slowly outward from the center of accumulation or down a mountain until it melts or breaks away

globe. any round, ball-shaped thing; a sphere; the Earth; a spherical model of the Earth showing features such as continents and seas

gorge. a deep, narrow pass between steep heights

grain. a small, hard seed or seedlike fruit, especially that of any cereal plant, as wheat, rice, corn, and rye

grassland. land with grass growing on it, used for grazing; pasture land

gravity. force that tends to draw all bodies in the Earth's sphere toward the center of the planet

greenhouse effect. the warming of the planet and its lower atmosphere caused by trapped solar radiation

gulf. a large area of ocean, larger than a bay, reaching into land

harbor. a protected inlet, or branch of a sea, lake, etc., where ships can anchor, especially one with port facilities

hemisphere. one of the planet's equal halves. The Equator divides the Earth into the northern and southern hemispheres; the Prime Meridian divides it into the eastern and western hemispheres.

hill. a natural raised part of the Earth's surface, often rounded and smaller than a mountain

horizon. the line where the sky seems to meet the Earth

humidity. the amount or degree of moisture in the air

humus. a brown or black substance resulting from the partial decay of plant and animal matter; the organic part of the soil

hurricane. a violent tropical cyclone with winds moving at 76 or more miles per hour, often accompanied by torrential rains, and often lethal in the West Indies

hydrosphere. all the water on the surface of the planet, including oceans, lakes, glaciers, etc., and including water vapor and clouds

Ice Age. any part of geologic time when large parts of the Earth were covered with glaciers; the latest of these times was the Pleistocene Epoch, when a large part of the northern hemisphere was intermittently covered with glaciers

iceberg. a great mass of ice broken off from a glacier and floating in the sea

ice cap. a mass of glacial ice that spreads slowly out in all directions from a center

ice sheet. a thick layer of ice covering an extensive area for a long period, as in an Ice Age

immigration. an act or instance of immigrating, coming into a new country, region, or environment, especially in order to settle there

irrigate. to refresh by or as by watering; to supply (land) with water by means of ditches or artificial channels

island. a land mass not as large as a continent, surrounded by water

isthmus. a narrow strip of land having water at each side and connecting two larger bodies of land

jungle. land covered with dense growth of trees, tall vegetation, vines, etc., typically in tropical regions and often termed a rain forest, and inhabited by predatory animals

karst. a region made up of porous limestone containing deep fissures and sinkholes and characterized by underground caves and streams

key. a reef or low island

lagoon. a shallow lake or pond, especially one connected with a larger body of water; the area of water enclosed by a circular coral reef, or atoll; an area of shallow salt water separated from the sea by sand dunes

lake. an inland body of water, usually fresh water, larger than a pool or pond, generally formed by some obstruction in the course of flowing water

landform. any topographic feature on the Earth's surface, as a plain, valley, or hill, caused by erosion, sedimentation, or movement

landscape. part of the planet's surface that can be viewed at one time from one place

landslide. the movement of a mass of loosened rocks or earth down a
  hillside or slope

latitude. the distance north or south of the Equator, measured in degrees

lava. melted rock issuing from a volcano

levee. an embankment built alongside a river to prevent high water
  from flooding adjacent land

lithosphere. the solid, rocky part of the Earth; Earth's crust

locate. to show the position of

loess. yellowish-brown, fine-grained soil deposited mainly by wind

longitude. the distance east or west of the Prime Meridian, represented
  by imaginary lines running north and south, measured in degrees

magma. liquid or molten rock deep in the Earth, which on cooling
  solidifies to produce igneous rock

magnetism. the force exerted by magnets when they attract other
  objects or repel each other

mantle. the layer of the Earth's interior between the crust and the core

map. a drawing or other representation, usually on a flat surface, of all
  or part of the Earth's surface, ordinarily showing countries, bodies of
  water, cities, mountains, etc.

marsh. a tract of low, wet, soft land that is either temporarily or
  permanently covered with water, characterized by aquatic, grasslike
  vegetation; a swamp; a bog; a morass; a fen

mesa. a small, high plateau or flat tableland with steep sides, especially
  in the southwest United States

meteorology. the science dealing with the atmosphere and especially
  the weather

migration. the movement from one place to another of a group of people,
  or of birds, fishes, etc.

mineral. an inorganic substance occurring naturally in the Earth and
  having a consistent and distinctive set of physical properties and a
  composition that can be expressed by a chemical formula; also,
  substances in the Earth of organic origin, such as coal

mining. the act, process, or work of removing ores, coal, etc., from a mine

mist. a large mass of water vapor at or just above the Earth's surface resembling a fog, but less dense

monsoon. a seasonal wind of the Indian Ocean and south Asia, blowing from the southwest from April to October, and from the northeast during the rest of the year

moraine. a mass of rocks, gravel, sand, clay, etc., carried and deposited directly by a glacier, along its side, at its lower end, or beneath the ice

mountain. a naturally raised part of the Earth's surface, usually rising more or less abruptly, and larger than a hill; a chain or group of such elevations

mouth. the point where a river enters a lake, a larger river, or the ocean

nation. a stable, historically developed community of people with a territory, economic life, distinctive culture, and language in common

natural resources. the actual and potential forms of wealth supplied by nature, such as coal, oil, water power, and arable land

navigation. the science of locating the position and plotting the course of ships and aircraft

neighborhood. a community, district, or area, especially with regard to some characteristic or point of reference

oasis. a fertile place in a desert, due to the presence of water

ocean. the great body of salt water that covers approximately 70 percent of the surface of the planet; any of its 4 principal geographical divisions: the Atlantic, Pacific, Indian, and Arctic Oceans

oceanography. the study of the environment in the oceans, including the waters, depths, beds, animals, and plants

oil. any of various kinds of greasy substances obtained from animal, vegetable, and mineral sources

ore. any natural combination of minerals, especially one from which a metal or metals can be profitably extracted

ozone layer. the atmospheric layer within the stratosphere that extends from a height of about 9 to about 18 miles and in which there is an appreciable concentration of ozone, responsible for absorbing much ultraviolet radiation and preventing some heat loss from the Earth

pass. a gap, or break, in high, rugged terrain such as a ridge in a mountain

peninsula. a land area almost entirely surrounded by water and connected with the mainland by an isthmus

permafrost. permanently frozen subsoil

petroleum. crude oil. Processed from petroleum are jet fuel, gasoline, and heating oil

piedmont. located at the base of a mountain or mountains, e.g., a piedmont stream, a piedmont area, or a piedmont plain, formed by soil moved by erosion down the mountain slope

plain. a large area of relatively flat land. Plains cover some 55 percent of the earth

plateau. an elevated tract of more or less level land; a tableland; a mesa

plate tectonics. the theory that the Earth's surface consists of plates, or large crustal slabs, whose constant motion explains phenomena such as continental drift and mountain building

pollute. to make unclean, impure, or corrupt; to contaminate

population. all the people in a country or region

port. a harbor; a city or town with a harbor where ships can load and unload cargo

prairie. a large area of level or slightly rolling grasslands, especially one in the Mississippi Valley

precipitation. rain, snow, sleet, etc., falling to the ground

Prime Meridian. the north-south line from which longitude is measured both east and west; 0° longitude, it passes through Greenwich, England

province. an administrative division of a country, such as the 10 main administrative divisions of Canada

race. a vague, unscientific term used to identify any of the different varieties or populations of human beings with distinguishing characteristics

rain. water falling in drops larger than 0.02 inches in diameter that have been condensed from the moisture in the atmosphere

rain forest. a dense, always green forest occupying a tropical region having abundant rainfall throughout the year

rain shadow. the dry lands that lie on the leeward side of mountains, the side away from the prevailing wind

rapids. part of a river where the current is relatively swift due to a narrowing of the river bed

region. a part of the surface of the Earth; a district; a division of the world characterized by a specific kind of plant or animal life; an area; a space

reservoir. a place where anything is collected and stored, generally in large quantity; especially, a natural or artificial lake or pond in which water is collected and stored for use

Richter scale. a logarithmic scale by which magnitude of earthquakes is measured, having graded steps, with each step approximately 10 times greater than the preceding step, and adjusted variously for different regions

Rift Valley. a depression of southwest Asia and east Africa, extending from the Jordan River valley across Ethiopia and Somalia to the lakes region of east Africa

Ring of Fire. a belt of volcanoes around the Pacific basin

river. a natural stream of water larger than a creek and emptying into an ocean, a lake, or another river

rock. a large mass of stone, the material that makes up most of the Earth; a natural substance, of either organic or inorganic origin, composed of solid mineral matter

rural. of or characteristic of the country, country life, or country people; having to do with farming; agricultural

savanna. a treeless plain or a grassland characterized by scattered trees, especially in tropical or subtropical regions having seasonal rains

sea. a division of the ocean that is enclosed or partly enclosed by land

sea level. the level of the surface of the ocean, especially the mean level between high and low tide; used as a standard in measuring heights, depths

season. any of the 4 arbitrary divisions of the year, characterized chiefly by differences in temperature, precipitation, amount of daylight, and plant growth: spring, summer, autumn, winter

sediment. matter deposited by water or wind

seismology. the geophysical science dealing with earthquakes and related phenomena

settlement. any place where people live

shelter. something that covers or protects; a protection, or place affording protection, as from the elements or danger

sierra. a range of hills or mountains having a saw-toothed appearance from a distance

silt. sediment suspended in stagnant water or carried by moving water, often accumulating on the bottom of rivers, bays, etc., especially sediment with particles smaller than sand and larger than clay

sinkhole. a saucer-shaped surface depression produced when underlying material, such as limestone or salt, dissolves or when caves, mines, etc., collapse

sleet. partly frozen rain, or rain that freezes as it falls

smog. a low-lying, perceptible layer of polluted air

snow. precipitation composed of ice crystals

soil. the layer of mineral and organic material that covers most of the Earth's land surface and supports plant life

solar pond. a shallow, stagnant pool with very hot, dense, salty water trapped at the lowest levels. This hot water can be converted into electricity or used directly

spring. a flow of water from the ground, often a source of a stream, pond, etc.

steppe. any of the great plains of southeast Europe and Asia, having few trees; any similar plain; a climatic term used to identify the transition regions between low-latitude deserts and humid climatic zones

strait. a narrow waterway connecting 2 large bodies of water

stream. a current or flow of water or other liquid, especially one running along the surface of the Earth; specifically, a small river or creek

swamp. a piece of wet, spongy land that is permanently or periodically covered with water, characterized by growths of shrubs and trees; a marsh; a bog

taiga. a transitional plant community that is located between the Arctic tundra and the boreal coniferous forests, having scattered trees

temperature. the degree of hotness or coldness of anything, usually as measured on a thermometer

terrace. a raised, flat mound of earth with sloping sides

terrain. ground or a tract of ground, especially with regard to its natural or topographical features or fitness for some use

terrane. a geological formation or series of related formations; a region where a specific rock or group of rocks predominates

tide. the alternate rise and fall of the surface of oceans and seas, and bays, rivers, etc., connected with them, caused by the attraction of the Moon and Sun. It may occur twice in each period of 24 hours and 50 minutes, which is the time of one rotation of the planet with respect to the Moon

timberline. the line above or beyond which trees do not grow, as on mountains or in polar regions

time zone. the time in any of the 24 zones, each an hour apart, into which the Earth is divided. It is based on distance east or west of the Prime Meridian, which passes through Greenwich, England

topography. the science of drawing on maps and charts or otherwise representing the surface features of a region, including its relief and rivers, lakes, etc., and such artificial features as canals, bridges, and roads

tornado. a violently whirling column of air, with wind speeds of about 100 to 300 miles per hour, extending downward from cumulonimbus clouds, especially in Australia and the central United States

town. an urban settlement generally larger than a village but smaller than a city

trade wind. a wind that blows steadily toward the Equator from the northeast in the tropics north of the Equator and from the southeast in the tropics south of the Equator

transpiration. the emission or exhalation of watery vapor from the surface of leaves or other parts of plants

tributary. a stream that flows into a larger one

tropics. the region that lies between the Tropic of Cancer, the line of latitude about 23.5° N of the Equator, and the Tropic of Capricorn, the line of latitude about 23.5° S of the Equator. The tropics encompass 36 percent of the Earth's land

tsunami. a huge sea wave caused by a great disturbance under an ocean, as a strong earthquake or volcanic eruption

tundra. any of the vast, nearly level, treeless plains of the arctic regions

typhoon. any violent tropical cyclone originating in the western Pacific Ocean, especially in the South China Sea

urban. referring to a city or town; characteristic of the city as distinguished from the country; citified; in US census use, designating or of an incorporated or unincorporated place with at least 50,000 inhabitants

valley. a stretch of low land lying between hills or mountains and usually having a river or stream flowing through it; the land drained or watered by a great river system

village. a group of houses in the country, larger than a hamlet and smaller than a city or town

volcano. a vent in the Earth's crust through which molten rock (lava), rock fragments, gases, ashes, etc., are ejected from the Earth's interior. A volcano is active while erupting, dormant during a long period of inactivity, or extinct when all activity has finally ceased

water. the colorless, transparent liquid occurring on Earth as rivers, lakes, oceans, etc., and falling from the clouds as rain; the most common substance on the planet and the only substance necessary to all forms of life. It exists in 3 forms: liquid, solid, gas

waterfall. a steep fall of water, as of a stream, from a height; a cascade

watershed. a ridge or stretch of highland dividing the areas drained by different rivers or river systems; the area drained by a river or river system

waterspout. a whirling funnel-shaped or tubelike column of air full of spray occurring over water, usually in tropical areas

water table. the level below which the ground is saturated with water

weather. the general condition of the atmosphere at a particular time and place, with regard to temperature, moisture, cloudiness, etc.

wetland. swamps or marshes; such an area preserved for wildlife

wilderness. an uncultivated, uninhabited region; any barren, empty, or open area

wind. the movement of air caused by the uneven heating of the planet by the Sun; a strong, fast-moving, or destructive natural current of air; a gale or storm

woodland. land covered with woods or trees; a forest

(Some of the definitions in this dictionary were drawn from *Webster's New World Dictionary of American English,* Third College Edition. Copyright © 1988 by Simon & Schuster, Inc., distributed by Prentice Hall Trade, 15 Columbus Circle, New York, New York 10023.)

The World's Least Developed Countries

The United Nations reported that these 41 "poorest countries" suffered even more degrading poverty, declining literacy, worsening health, and generally lower living standards in the 1980s—all made worse by large population increases, fragile environments, vulnerability to natural disasters, and, often, limited civil rights or outright repression:

Afghanistan	Malawi
Bangladesh	Maldives
Benin	Mali
Bhutan	Mauritania
Botswana	Mozambique
Burkina Faso	Myanmar (Burma)
Burundi	Nepal
Cape Verde	Niger
Central African Republic	Rwanda
Chad	Samoa
Comoros	Sao Tome and Principe
Djibouti	Sierra Leone
Equatorial Guinea	Somalia
Ethiopia	Sudan
Gambia	Tanzania
Guinea	Togo
Guinea-Bissau	Tuvalu
Haiti	Uganda
Kiribati	Vanuatu
Laos	Yemen
Lesotho	

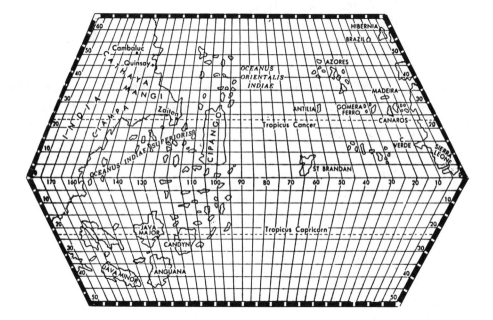

1. Myanmar is the new name of
    a. Burma
    b. Kenya
    c. Papua New Guinea
    d. Uganda

2. Only 2 South American countries do not have a common border with Brazil. They are
    a. Ecuador and Chile
    b. Paraguay and Uruguay
    c. Peru and Colombia
    d. Chile and Argentina

3. _____ countries extend north of the Arctic Circle, the parallel at 66.5° N, which separates the Arctic region from the North Temperate Zone.
    a. Two
    b. Four
    c. Seven
    d. Eleven

4. Guatemala, in Central America,
    a. has 3 large deserts
    b. has more than 30 volcanoes
    c. is the flattest country between Canada and South America
    d. has never had an earthquake

5. Name 2 of the 4 African countries that are part of the Maghrib:
    a. Benin and Ghana
    b. Mali and Zaire
    c. Libya and Tunisia
    d. Egypt and Chad

6. The smallest and most densely populated of the republics of Central America is
   a. El Salvador
   b. Panama
   c. Honduras
   d. Guatemala

7. When Southern Rhodesia's black population—there were at least 9 blacks to every white—took control of the government in 1980, they renamed their African nation
   a. Namibia
   b. Zimbabwe
   c. Uganda
   d. Cameroon

8. The only country in Central America that isn't on the Atlantic coast is
   a. Costa Rica
   b. El Salvador
   c. Belize
   d. Guatemala

9. A 1,000-mile-long fence was built in _____ to keep rabbits from overrunning the entire country.
   a. Mongolia
   b. Australia
   c. South Africa
   d. Chile

10. In 1971, East Pakistan became
    a. Singapore
    b. Bangladesh
    c. Sri Lanka
    d. Sumatra

11. The 2 landlocked countries of South America are
    a. Bolivia and Paraguay
    b. Paraguay and Uruguay
    c. Colombia and Peru
    d. Bolivia and French Guiana

12. The 2 countries that share the Iberian Peninsula are
    a. Australia and New Guinea
    b. Korea and Mongolia
    c. Mexico and the United States
    d. Spain and Portugal

13. Almost all of the world's cocaine is produced in
    a. Iran
    b. Thailand
    c. Colombia
    d. India

14. Mexico trades more with _____ than with any other country.
    a. Japan
    b. the Netherlands
    c. the United States
    d. Brazil

15. The smallest republic in the world is
    a. Monaco
    b. the Holy See
    c. San Marino
    d. Andorra

16. Canada has _____ provinces.
    a. 5
    b. 6
    c. 8
    d. 10

17. The Belgian Congo, in Africa, gained its independence in 1960 and is now called
    a. Cameroon
    b. Zaire
    c. Mauritania
    d. Rwanda

18. The largest Scandinavian country in square miles is
    a. Denmark
    b. Sweden
    c. Finland
    d. Norway

19. The world's leading producer of coffee is
    a. Mali
    b. Java
    c. Brazil
    d. Gabon

20. Both Pakistan and Brazil
    a. built new capital cities
    b. exported their homeless to neighboring countries
    c. have more than 20 nuclear power plants
    d. do not permit the showing of motion pictures

21. The world's first urban nation was
   a. Great Britain
   b. Germany
   c. the United States
   d. India

22. If you were on the road to Mandalay, as the song says, you'd be going to
   a. Myanmar
   b. Natal
   c. the Hawaiian Islands
   d. Singapore

23. Zambia, home to more than 70 ethnic groups, in south-central Africa, is bordered by _____ other nations.
   a. 5
   b. 8
   c. 9
   d. 12

24. _____ is the country with the highest greenhouse-gas net emissions.
   a. The United States
   b. Japan
   c. Saudi Arabia
   d. Poland

25. More than 50 percent of the automobiles in the South American country of _____ run on gasohol.
   a. Colombia
   b. Venezuela
   c. Brazil
   d. Argentina

26. Two South American countries are members of OPEC—the Organization of Petroleum Exporting Countries. The countries are
    a. Ecuador and Venezuela
    b. Brazil and Venezuela
    c. Colombia and Ecuador
    d. Argentina and Chile

27. In 1957, British Togoland joined the Gold Coast and nearby British-administered territories and they became the independent nation of
    a. Ghana
    b. Zanzibar
    c. Tanganyika
    d. Tanzania

28. The Sudan is the largest country in Africa. The second largest is
    a. Algeria
    b. Egypt
    c. Angola
    d. the Republic of South Africa

29. The United Arab Emirates, on the east Arabian Peninsula, were once called
    a. Trucial States
    b. the Seven Sheikhdoms
    c. Solomon's Mines
    d. the Dubai States

30. Switzerland has _____ official languages.
    a. 2
    b. 3
    c. 4
    d. 6

31. Germany was formally divided into 2 republics, West Germany and East Germany, in
    a. 1871
    b. 1919
    c. 1933
    d. 1949

32. The world's largest exporter of electricity is
    a. the Soviet Union
    b. Canada
    c. Paraguay
    d. Switzerland

33. The only African country with a coastline along both the Atlantic and Pacific Oceans is
    a. Kenya
    b. the Republic of South Africa
    c. Morocco
    d. Sudan

34. Mozambique is to Portugal what Kenya is to
    a. Spain
    b. the United Kingdom
    c. Belgium
    d. the Benelux countries

35. The main portion of west Africa's Gold Coast—so named for the huge amounts of gold hauled away from its sands and mines—is today called
    a. Kuwait
    b. Ghana
    c. Uganda
    d. Senegal

36. The only country to contain both the Equator and a tropic is
    a. Brazil
    b. Borneo
    c. Australia
    d. Peru

37. This country has the highest population-growth rate in the world:
    a. Kenya
    b. China
    c. Mexico
    d. India

38. Monaco, the second smallest independent state in the world, is about the size of
    a. New York City's Central Park
    b. the Astrodome, in Houston, Texas
    c. Las Vegas, Nevada
    d. Lake Huron

39. In 1667 the British traded Suriname in South America to the Netherlands for the Dutch colony of
    a. Java
    b. New Amsterdam
    c. Celebes
    d. Sumatra

40. This country's neutrality was guaranteed in 1815 by the Congress of Vienna:
    a. Switzerland's
    b. Liechtenstein's
    c. Luxembourg's
    d. Denmark's

41. _____ countries have more land area than the United States.
   a. Two
   b. Three
   c. Six
   d. Eleven

42. Only 1 black African country escaped colonialism:
   a. Liberia
   b. Senegal
   c. Gambia
   d. Zaire

43. The Soviet Union annexed the Baltic States in
   a. 1789
   b. 1812
   c. 1919
   d. 1940

44. The Soviet Union has _____ time zones.
   a. 5
   b. 7
   c. 10
   d. 11

45. One of these nations does not border Mexico:
   a. Belize
   b. Guatemala
   c. the United States
   d. El Salvador

46. This colony in central-south Africa was divided into 2 new nations:
  a. Rhodesia
  b. Bechuanaland
  c. Angola
  d. Namibia

47. The world's oldest black republic is
  a. Haiti
  b. Panama
  c. the Central African Republic
  d. Burundi

48. The west African nation of Senegal is penetrated for 200 miles from the Atlantic coast by the nation of
  a. Gambia
  b. Togo
  c. Sierra Leone
  d. Gabon

49. One of the following 4 countries is not among the 3 largest countries in the southern hemisphere:
  a. Brazil
  b. Australia
  c. Argentina
  d. South Africa

50. Patagonia, a region of about 311,000 square miles at the southern tip of South America, is shared by Argentina and
  a. Great Britain
  b. Uruguay
  c. New Zealand
  d. Chile

51. Four newly industrialized Asian countries with rising economies are
     Singapore, Hong Kong, Taiwan, and South Korea. They are
     popularly referred to as
     a. the Four Tigers
     b. the Asian Quartet
     c. the Bolts from the Blue
     d. the Disrupters

52. The cradle of Western civilization was the country whose name
     meant "between rivers." The country was
     a. India
     b. Afghanistan
     c. Mesopotamia
     d. Egypt

53. The western European countries of France and Spain are separated by
     a. the Po River
     b. the Pyrenees Mountains
     c. the Alps
     d. the Atlas Mountains

54. The African countries of Gabon, Chad, Congo, and the Central
     African Republic were once part of
     a. the Belgian Congo
     b. Nyasaland
     c. French Equatorial Africa
     d. Egypt

55. When the People's Republic of China was admitted into the United
     Nations in 1971, this country was forced to withdraw from the world
     body:
     a. Japan   **c.** Taiwan
     b. Korea   **d.** Cambodia

56. Many homes are heated with hot water pumped directly from the ground in
   a. Gambia
   b. Western Sahara
   c. Iceland
   d. Honduras

57. This nation was the first to begin taking a census of its population every decade:
   a. France
   b. Italy
   c. India
   d. the United States

58. Nuclear submarines are not allowed to dock in
   a. Greece
   b. Argentina
   c. Sri Lanka
   d. New Zealand

59. This northern European nation, which is about the size of the US states of New Jersey and New York and the 6 New England states combined, has more than 60,000 lakes and is more than 70 percent forested:
   a. Norway
   b. Sweden
   c. Finland
   d. Poland

60. Peru has 2 official languages. One is Quechua. The other is
   a. Portugese   **c.** English
   b. Spanish   **d.** Incan

61. The most dangerous and frequent natural hazard in Bangladesh is
    a. flooding
    b. volcanic eruptions
    c. earthquakes
    d. blizzards

62. The greatest financial assistance to less developed countries is provided by
    a. the United States
    b. the United Kingdom
    c. Japan
    d. France

63. "The Great Lone Land" is the popular nickname for
    a. Antarctica
    b. the Arctic
    c. Canada
    d. South Africa

64. Ben Nevis is Great Britain's
    a. highest peak
    b. largest lake
    c. longest river
    d. smallest hamlet

65. These countries were established in this century primarily for religious reasons:
    a. Pakistan and Israel
    b. Kenya and Zaire
    c. Iraq and Iran
    d. Singapore and Sri Lanka

66. Dahomey, in west Africa, assumed its new name of _____
    in 1976.
    a. Senegal
    b. Ghana
    c. Benin
    d. Gabon

67. The African nation whose cities of Monrovia and Buchanan honor
    Presidents of the United States is
    a. Nigeria
    b. Niger
    c. Zaire
    d. Liberia

68. Iceland's principal export is
    a. mineral water
    b. volcanic ash
    c. fish
    d. ship-building acumen

69. The southwest Pacific countries of Australia and New Zealand are
    about _____ miles from each other.
    a. 100
    b. 400
    c. 1,200
    d. 7,500

70. One of these countries is not on 0° longitude:
    a. England
    b. South Africa
    c. Mali
    d. Spain

71. One of these countries is not on 60° North latitude:
     a. the Soviet Union
     b. Canada
     c. Sweden
     d. Denmark

72. One of these countries is not on the Equator:
     a. Kenya
     b. Australia
     c. Brazil
     d. Indonesia

73. French Guiana, on the northeast coast of South America, was famous for
     a. vast reserves of bauxite
     b. massive, carved stone heads made by persons unknown
     c. its prison colony
     d. its 5 rain forests

74. War, like earthquakes, dramatically alters geography. During the Vietnam war, US airplanes reshaped this country with 2,093,100 tons of bombs:
     a. North Vietnam
     b. Cambodia
     c. Laos
     d. Thailand

1. **(a)** Myanmar, pronounced mee-ahn-MAH, is the new name of Burma, in southeast Asia. The name of the capital city has also been changed, from Rangoon to Yangon. The new names reflect contemporary usage in the "Burmese" language. Many place names in the language had been adapted into English during British colonial rule there, between 1862 and 1948.

2. **(a)** Ecuador and Chile are the only 2 South American countries that do not have a common border with Brazil.

3. **(c)** The 7 countries that extend north of the Arctic Circle are Canada, Finland, Iceland, Norway, the Soviet Union, the United States, and Sweden.

4. **(b)** There are more than 30 volcanoes in Guatemala.

5. **(c)** Libya, Tunisia, Morocco, and Algeria are part of the Maghrib, the Arabic name for northwest Africa. During the Moorish occupation, the Maghrib included most of Spain.

6. **(a)** El Salvador, 8,260 square miles, with a population of about 5 million, is the smallest and most densely populated republic in Central America.

7. **(b)** Zimbabwe became the new name for Southern Rhodesia when the majority blacks took control of the government in 1980.

8. **(b)** El Salvador is the only Central American country without an Atlantic seaboard.

9. **(b)** To keep rabbits from overrunning the entire country, a 1,000-mile-long fence was put up in the middle of Australia.

10. **(b)** Bangladesh, an independent nation and the "world's basket case," became the name for East Pakistan in 1971.

11. **(a)** Bolivia and Paraguay are the only 2 countries in South America that do not border on an ocean.

12. **(d)** Spain and Portugal are the 2 countries on the Iberian Peninsula in southwest Europe.

13. **(c)** The South American country of Colombia produces almost all of the world's cocaine.

14. **(c)** Mexico's number-one trading partner is the United States.

15. **(c)** San Marino, 24 square miles, on Mount Titano in central Italy, is the oldest state in Europe and the world's smallest republic. The population is around 20,000. The government was communist from 1945 to 1957.

16. **(d)** The 10 Canadian provinces are Alberta, British Columbia, Manitoba, New Brunswick, Newfoundland, Nova Scotia, Ontario, Prince Edward Island, Quebec, and Saskatchewan.

17. **(b)** Zaire is the name of the former Belgian Congo.

18. **(b)** Sweden, 173,665 square miles, is the largest Scandinavian country. Norway is 125,049 square miles; Denmark, 16,629 square miles; Finland (which is not part of Scandinavia), 117,944 square miles.

19. **(c)** Brazil is the world's leading producer of coffee.

20. **(a)** Pakistan built the new capital city of Islamabad, and Brazil built Brasilia; both new cities went up in sparsely settled areas.

21. **(a)** A country where most of the people live in cities is described as an urban nation. The world's first urban nation was Great Britain, after industrialization in the late 1800s.

22. **(a)** You'd be going to Myanmar (formerly known as Burma) if you were on the road to Mandalay, a religious center for Buddhists about 365 miles north of Yangon. The city was largely destroyed by the occupying Japanese military during the Second World War.

23. **(b)** The south-central African country of Zambia is bordered by 8 other countries: Angola, Botswana, Malawi, Mozambique, Namibia, Tanzania, Zaire, and Zimbabwe.

24. **(a)** In order of their index rank, the United States emits 17.6 percent of the world's greenhouse gas; the Soviet Union, 12 percent; Brazil, 10.5; China, 6.6; Japan and India, each 3.9.

25. **(c)** More than 50 percent of the automobiles in Brazil run on gasohol.

26. **(a)** Ecuador and Venezuela are the South American members of OPEC.

27. **(a)** Ghana, in west Africa, became independent in 1957. (In east Africa, Tanzania is the union of Zanzibar and Tanganyika.)

28. **(a)** Algeria, which is about one third the size of the continental United States, is the second largest country in Africa. Ninety-one percent of the population lives along its Mediterranean coastline of about 620 miles.

29. **(a)** The United Arab Emirates were formerly known as Trucial States, or Trucial Oman, or Trucial Coast.

30. **(c)** The 4 official languages of Switzerland are German, French, Italian, and Romansh. Few Swiss speak Romansh.

31. **(d)** West Germany and East Germany, both republics, were constituted in 1949. (Four decades later, Germany became one country again.)

32. **(c)** The world's largest exporter of electricity is the landlocked South American country of Paraguay. It jointly developed massive hydroelectric projects on the Paraná River with Argentina and Brazil, and it sells to its larger partners much of its share of the electricity generated.

33. **(b)** The Republic of South Africa is the only African country that borders on both the Atlantic and the Pacific Oceans.

34. **(b)** Mozambique is to Portugal what Kenya is to the United Kingdom. Mozambique was a Portuguese colony; Kenya was a UK colony.

35. **(b)** The Gold Coast, most of which became the republic of Ghana, on the west coast of Africa, was a center for the slave trade and known for its gold.

36. **(a)** Brazil is the only country in the world to contain both the Equator and a tropic (the Tropic of Capricorn).

37. **(a)** The population of the east African country of Kenya increases at the rate of about 4 percent a year, the highest growth rate in the world. By the year 2020, its population is expected to be about 46 million. The United States, by contrast, has a fairly low population-growth rate of less than 1 percent annually.

38. **(a)** Monaco—0.8 square miles—is about the size of Central Park in New York City. The principality, which is second only to Vatican City as the smallest independent state in the world, is on the Mediterranean coast 11 miles from Nice, France, and has been ruled by the House of Grimaldi since 1419. It was founded in 1215 as a colony of Genoa.

39. **(b)** In 1667 the British traded Suriname in South America for the Netherlands' New Amsterdam, present-day Manhattan, New York.

40. **(a)** Switzerland's neutrality was guaranteed by the Congress of Vienna in 1815. The country remained neutral even in both world wars.

41. **(b)** Three countries—the Soviet Union, Canada, and China—each have more land area than the United States.

42. **(a)** Liberia is the only black African country to escape colonialism. In 1847 it became the continent's first independent republic, then spent the next century, its "century of survival," warding off encroachment by France and Great Britain, both colonial powers.

43. **(d)** The Soviet Union annexed the 3 Baltic States—Estonia, Latvia, and Lithuania—in 1940 in the wake of its nonaggression pact with Hitler's Germany.

44. **(d)** The Soviet Union has 11 time zones. When it is 6 PM on the Pacific coast, it is 7 AM in Leningrad on the nation's western border.

45. **(d)** Belize, Guatemala, and the United States border on Mexico; El Salvador does not.

46. **(a)** Rhodesia—in a region that was the home of humans since earliest times—was divided into Zimbabwe (Southern Rhodesia) and Zambia (Northern Rhodesia).

47. **(a)** Haiti, in the Caribbean Sea, is the world's oldest black republic, independent since 1802.

48. **(a)** Gambia penetrates 200 miles into Senegal, on the west coast of Africa.

49. **(d)** Brazil is the largest country in the southern hemisphere; Australia is the second largest; Argentina, the third.

50. **(d)** In 1881 Patagonia was divided between Argentina and Chile.

51. **(a)** The Four Tigers are newly industrialized Asian countries with rising economies: Singapore, Hong Kong, Taiwan, and South Korea. They are also called the Little Dragons.

52. **(c)** Mesopotamia, literally "between rivers," namely the Tigris and the Euphrates, had land-cultivating settlements as early as 5000 BC and soon thereafter had city states such as Ur and Erech. Mesopotamia was where Iraq is today.

53. **(b)** The Pyrenees Mountains separate France and Spain.

54. **(c)** Gabon, Chad, the Congo, and the Central African Republic were once part of French Equatorial Africa.

55. **(c)** Taiwan, the island seat of the Chinese nationalist government, was forced to withdraw from the United Nations when the world organization admitted the People's Republic of China in 1971.

56. **(c)** Many homes in Iceland are heated with hot water pumped directly from the ground.

57. **(d)** The United States was the first nation to begin taking a census of its population every 10 years. Article I, Section 2 of the Constitution states: "The actual enumeration shall be made within three years after the first meeting of the Congress of the United States, and within every subsequent term of ten years, in such manner as they by law direct." The first US census was in 1790, 2 years after ratification of the Constitution.

58. **(d)** By declaring itself a nuclear-free zone, New Zealand denies docking rights to nuclear submarines.

59. **(c)** Low but hilly Finland has more than 60,000 lakes and is more than 70 percent forested.

60. **(b)** Quechua and Spanish are the 2 official languages of Peru.

61. **(a)** Flooding is the most dangerous and frequent natural hazard in Bangladesh, in southern Asia. Deforestation in the Himalayas—massive cutting of firewood is the major cause of the deforestation—contributes to the Bangladesh floods.

62. **(c)** Japan provides the greatest financial assistance to less developed countries.

63. **(c)** Canada, the second largest country in the world, is "The Great Lone Land." A million square miles is tundra.

64. **(a)** Ben Nevis, 4,406 feet, is the highest mountain in Great Britain. It is in west-central Scotland, east of the northern end of Loch Linnhe.

65. **(a)** Both Pakistan and Israel were established primarily for religious reasons.

66. **(c)** Dahomey became the west African republic of Benin in 1976.

67. **(d)** Monrovia, the capital of Liberia, was named for US President James Monroe, and the city of Buchanan was named for US President James Buchanan.

68. **(c)** Fish is the principal export of the North Atlantic island of Iceland.

69. **(c)** Australia and New Zealand in the southwest Pacific Ocean are about 1,200 miles apart.

70. **(b)** England, Mali, and Spain are on 0° longitude; South Africa is not.

71. **(d)** The Soviet Union, Canada, and Sweden are on 60° North latitude; Denmark is not.

72. **(b)** Kenya, Brazil, and Indonesia are on the Equator; Australia is not.

73. **(c)** French Guiana was famous for its prison colony on Devils Island and for its prison camps at Kourou and Saint-Laurent on the mainland.

74. **(c)** America's "secret war" in southeast Asia to interdict the Ho Chi Minh Trail, North Vietnam's supply route through neighboring Laos during the Vietnam war, went on for 9 years. By 1973, when the bombing of Laos stopped, the US had flown 580,944 sorties over the country, unleashing a third more tonnage than had pulverized Nazi Germany in the Second World War and three times the tonnage dropped during the Korean war.

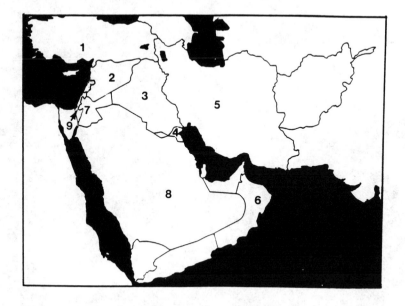

MQ1. Match the 9 labeled countries of the Middle East:

Saudi Arabia	**1.**	_____
Israel	**2.**	_____
Syria	**3.**	_____
Turkey	**4.**	_____
Kuwait	**5.**	_____
Iraq	**6.**	_____
Oman	**7.**	_____
Jordan	**8.**	_____
Iran	**9.**	_____

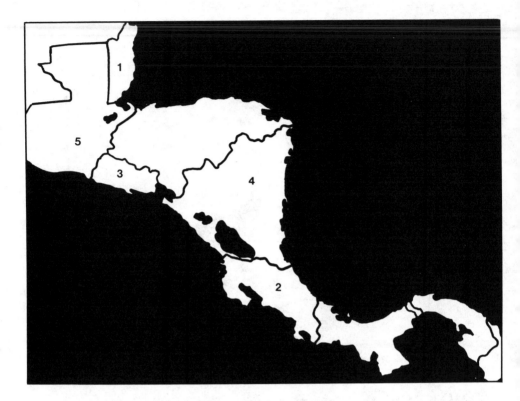

MQ2. Match the 5 labeled countries of Central America:

Belize	1.	_____
El Salvador	2.	_____
Guatemala	3.	_____
Nicaragua	4.	_____
Costa Rica	5.	_____

MQ3. Match the 7 labeled countries of South America:

Uruguay	1. _____
Paraguay	2. _____
Guyana	3. _____
Bolivia	4. _____
Ecuador	5. _____
French Guiana	6. _____
Suriname	7. _____

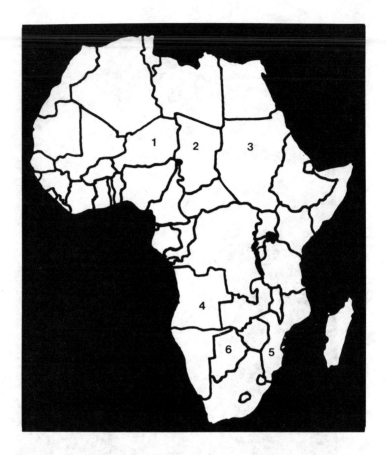

MQ4. Match the 6 labeled countries of Africa:

Chad   **1.** _____

Botswana   **2.** _____

Niger   **3.** _____

Mozambique   **4.** _____

Sudan   **5.** _____

Angola   **6.** _____

MQ5. Match the 6 labeled countries of Asia:

Nepal	1.	_____
Malaya	2.	_____
Borneo	3.	_____
Bangladesh	4.	_____
Mongolia	5.	_____
Thailand	6.	_____

MQ6. Match the 7 labeled countries:

Turkey	1. _____
the Netherlands	2. _____
Ireland	3. _____
Portugal	4. _____
Sweden	5. _____
Romania	6. _____
Czechoslovakia	7. _____

MQ7. Match the 6 labeled provinces and 1 territory of Canada:

Yukon Territory

Saskatchewan

Nova Scotia

British Columbia

Manitoba

New Brunswick

Newfoundland

1. _____

2. _____

3. _____

4. _____

5. _____

6. _____

7. _____

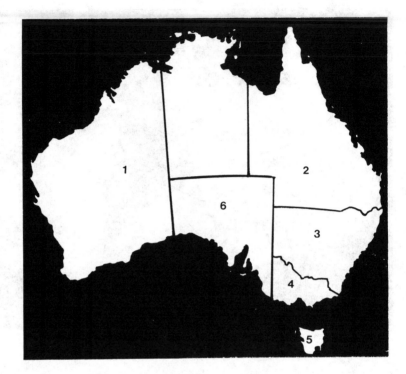

MQ8. Match the 6 labeled states of Australia:

Tasmania	**1.** _____
New South Wales	**2.** _____
Western Australia	**3.** _____
South Australia	**4.** _____
Queensland	**5.** _____
Victoria	**6.** _____

MQ9. This 3,284,426-square-mile country has a 4,603-mile-long coastline with only a couple of large offshore islands:

    a. France

    b. the People's Republic of China

    c. Brazil

    d. Norway

MQ10. Place an X on the Nevada-size east Asian nation in which, according to the World Bank, half of the 60 million population live in "absolute poverty," their income unable "to satisfy basic needs."

MQ11. This 2,650-mile-long country has several mountain peaks above 19,000 feet and was for years a chief source of the world's nitrates:

   a. Nepal
   b. Senegal
   c. Chile
   d. Tibet

MQ12. This is a map of the northeast African republic of

   a. Chad
   b. Libya
   c. Egypt
   d. Ethiopia

MQ13. These 2 European countries are:

Poland	Germany
Greece	Switzerland
Spain	Italy
France	Austria

1. _____

2. _____

MQ14. This is a map of

a. Albania

b. Laos

c. Vermont

d. the Congo

MQ15. This is a map of the northwest African kingdom of

a. Algeria

b. Morocco

c. Tunisia

d. Western Sahara

MQ16. This is a map of the central European nation of
    a. Romania
    b. Bulgaria
    c. Yugoslavia
    d. Poland

MQ17. This is a map of
    a. Vietnam
    b. Panama
    c. North and South Korea
    d. Florida

MQ18. This is a map of _____, the largest

island in Europe, and the 4 labeled regions are:

1. _____

2. _____

3. _____

4. _____

MQ19. The 5 labeled countries in the western hemisphere are:

Peru	1. _____
Venezuela	2. _____
Mexico	3. _____
Argentina	4. _____
Brazil	5. _____

MQ1. The 9 labeled countries of the Middle East are: 1. Turkey; 2. Syria; 3. Iraq; 4. Kuwait; 5. Iran; 6. Oman; 7. Jordan; 8. Saudi Arabia; 9. Israel.

MQ2. The 5 labeled countries of Central America are: 1. Belize; 2. Costa Rica; 3. El Salvador; 4. Nicaragua; 5. Guatemala.

MQ3. The 7 labeled countries of South America are: 1. Guyana; 2. Suriname; 3. French Guiana; 4. Ecuador; 5. Uruguay; 6. Paraguay; 7. Bolivia.

MQ4. The 6 labeled African countries are: 1. Niger; 2. Chad; 3. Sudan; 4. Angola; 5. Mozambique; 6. Botswana.

MQ5. The 6 labeled Asian countries are: 1. Borneo; 2. Mongolia; 3. Thailand; 4. Nepal; 5. Bangladesh; 6. Malaya.

MQ6. The 7 labeled countries are: 1. Portugal; 2. the Netherlands; 3. Czechoslovakia; 4. Sweden; 5. Romania; 6. Turkey; 7. Ireland.

MQ7. The 6 labeled provinces and 1 territory of Canada are: 1. British Columbia; 2. Saskatchewan; 3. Manitoba; 4. Newfoundland; 5. New Brunswick; 6. Nova Scotia; 7. Yukon Territory.

MQ8. The 6 labeled states of Australia are: 1. Western Australia; 2. Queensland; 3. New South Wales; 4. Victoria; 5. Tasmania; 6. South Australia. (The unnumbered region in the northern center of the continent is a territory, the Northern.)

MQ9. (c) Brazil, the largest country in South America, is larger than the contiguous 48 US states. Its heavily-wooded Amazon basin covers half the country. Almost half the population lives in the south-central region, which produces 80 percent of the nation's industrial output and 75 percent of its farm products.

MQ10. Thirty million people live in absolute poverty in the Philippines, an archipelago of about 7,100 islands approximately 500 miles off the southeast coast of Asia. About 95 percent of the population lives on the 11 largest islands.

MQ11. (c) The republic of Chile on the western coast of southern South America is nowhere more than 221 miles wide and it has no rivers of size. It has rarely rained, much less sprinkled, in the Atacama Desert, in the north-central part of the country.

MQ12. (c) Egypt's 386,900 square miles (about the size of Arkansas, Texas, and Oklahoma combined) are bounded by the Mediterranean Sea, Israel, the Red Sea, the Sudan, and Libya. The ancient country is almost entirely desolate and barren; only 4 percent of the land is arable. Almost all Egyptians (more than 50 million) live in the 550-mile-long Nile Valley.

MQ13. The 2 European countries are (1) Spain, 194,881 square miles, occupying the greater part of the Iberian Peninsula in southwestern Europe and about the size of Utah and Arizona combined, and (2) Greece, 50,944 square miles, about the size of Alabama, in the mountainous southern end of the Balkan Peninsula in southeastern Europe; about 75 percent of Greece is non-arable.

MQ14. **(a)** Albania, which is slightly larger than the state of Maryland, is between Yugoslavia and Greece on the east coast of the Adriatic Sea. Its 11,100 square miles are very mountainous. The Communist nation is officially atheistic; all public worship and religious instruction were outlawed in 1967.

MQ15. **(b)** Morocco's 172,413 square miles (bigger than California), in northwest Africa, are bounded by the Mediterranean Sea, Algeria, Western Sahara, and the Atlantic Ocean. Stretching from the southwest to the northeast are the Atlas Mountains; the highest point is Toubkal, 13,671 feet. To the west of southern Morocco are the Canary Islands.

MQ16. **(d)** Poland, 120,756 square miles, is bounded by the Baltic Sea and the Soviet Union, by Czechoslovakia and Germany. Its northern and central regions are mostly lowlands. The Carpathian Mountains along the southern border rise to 8,200 feet.

MQ17. **(a)** The southeast Asian Socialist Republic of Vietnam, which is about the size of New Mexico, was divided into North Vietnam and South Vietnam in the period 1954–75, then was united when the South Vietnamese government fell. The coastline is 1,400 miles.

MQ18. The 4 labeled regions of the United Kingdom, the largest island in Europe, are: 1. Northern Ireland; 2. Wales; 3. Scotland; 4. England.

MQ19. The 5 labeled countries in the western hemisphere are: 1. Brazil; 2. Mexico; 3. Peru; 4. Argentina; 5. Venezuela.

1. The oldest rocks in the world are in
   a. Antarctica
   b. Canada
   c. India
   d. Switzerland

2. The land somewhere in Japan rocks with an earthquake on the average of
   a. once a week
   b. 4 times a day
   c. once a month
   d. 6 times a year

3. The Mariana Trench is
   a. the deepest place in the oceans of the world
   b. the continental plate that pushed up the Himalayas
   c. the continental plate on which Australia is floating around
   d. a canyon in Siberia that is second in size only to Hells Canyon in the northwest United States

4. Ice covers _____ of the land surface of the Earth.
   a. a third of 1 percent
   b. 2.5 percent
   c. 10 percent
   d. nearly 17 percent

5. There are more than 100 volcanoes and about 120 glaciers in
   a. Alaska
   b. the Soviet Union
   c. Central America
   d. Iceland

6. The world's largest glacier—260 miles long—is in

    a. Canada

    b. Alaska

    c. New Zealand

    d. Antarctica

7. Mount Erebus is a volcano in

    a. Alaska

    b. Japan

    c. Angola

    d. Antarctica

8. The most recent ice age began retreating _____ years ago.

    a. 18,000

    b. 180,000

    c. 1,800,000

    d. nearly 10,000,000

9. The planet has 4 oceans and _____ seas.

    a. 10

    b. 32

    c. 39

    d. 83

10. The largest hole in the Earth's vital ozone layer is located over

    a. Brazil

    b. the Equator

    c. Antarctica

    d. the Indian Ocean

11. The highest active volcano in the world is in
     a. Japan
     b. Alaska
     c. Indonesia
     d. Chile

12. A volcano erupts when hot, molten rock, or magma, rises from the Earth's mantle to its surface. Cone-shaped Mount Saint Helens erupted on May 18, 1980, in the US state of
     a. Oregon
     b. New Mexico
     c. California
     d. Washington

13. The largest depressions on the planet are
     a. canyons
     b. ocean basins
     c. bayous
     d. dead seas

14. The solid outer part of the Earth is the
     a. lithosphere
     b. atmosphere
     c. hydrosphere
     d. biosphere

15. The Ring of Fire is
     a. a line of active volcanoes encircling the basin of the Pacific Ocean
     b. a nickname for the Hawaiian Islands
     c. the crater of Africa's largest volcano
     d. a national park on the Equator in Africa

16. Prime Meridian—0° longitude—is located at
    a. the peak of Mount Everest
    b. Warsaw, Poland
    c. Buenos Aires, Argentina
    d. Greenwich, England

17. Four great desert regions—the Great Sandy, the Outback, the Gibson, and the Great Victoria—are in
    a. northern Africa
    b. southern Africa
    c. Australia
    d. northeastern Siberia

18. The desert along the southwest coast of Africa is noted for
    a. aridity
    b. the mining of diamonds
    c. violent thunderstorms, a minimum of 3 a day
    d. its unique animals

19. There are about _____ active volcanoes in the world.
    a. 12
    b. 28
    c. 361
    d. 600

20. About _____ percent of the Earth's surface is covered with water.
    a. 12
    b. 23
    c. 53
    d. 70

21. The _____ is exactly 24,901.5 miles.
    a. distance from Washington, D.C., to Moscow, the Soviet Union,
    b. length of the Equator
    c. width of the Pacific Ocean
    d. length of the International Date Line

22. In the year _____ , the supersonic transport the Concorde began cutting flight time across the Atlantic Ocean in half.
    a. 1953
    b. 1959
    c. 1967
    d. 1976

23. The deepest chasm in North America is
    a. Hells Canyon
    b. the Grand Canyon
    c. Black Canyon
    d. Ausable Chasm

24. A desert is an area of land that receives less than _____ inches of precipitation a year.
    a. 30
    b. 22
    c. 12.5
    d. 10

25. The collision of the Eurasian and Indo-Australian tectonic plates gave birth to
    a. Antarctica
    b. the Philippines
    c. the world's highest mountains
    d. Africa

26. The 51-mile-long Dead Sea, actually a salt lake, is the lowest point on the surface of the planet, and is shared by Israel and
    a. Jordan
    b. Syria
    c. Lebanon
    d. Egypt

27. The greatest known depth of the Indian Ocean is nearly
    a. 2 miles
    b. 3 miles
    c. 4 miles
    d. 5 miles

28. To the nearest mile, _____ miles separate the length of 1 degree of latitude from the next.
    a. 14 or 15
    b. 29 or 30
    c. 49 or 50
    d. 69 or 70

29. "Vog" is short for
    a. coke residue
    b. geyser overflow
    c. ozone hole
    d. volcanic fog

30. The largest glacier in North America is in
    a. Hudson Bay
    b. the Northwest Territories
    c. Alaska
    d. Alberta

31. The longest lowest gap in the western hemisphere is in
    a. Mexico
    b. Nicaragua
    c. Panama
    d. Colombia

32. The fine, silty sediment deposited by wind is called
    a. loess
    b. haboob
    c. haar
    d. ubac

33. Mexico's highest point is the volcanic peak
    a. Popocatepetl
    b. Citlaltepetl
    c. Nevado de Toluca
    d. Colima

34. Northeast Sicily's active volcano, Etna, has had about _____ eruptions.
    a. 11
    b. 42
    c. 55
    d. 140

35. Two of North America's best known volcanoes are side by side in the central region of
    a. Alaska
    b. Oregon
    c. Greenland
    d. Mexico

1. **(b)** Chunks of granite 4 billion years old have been found south of the Arctic Circle in the Northwest Territories, in remote northern Canada.

2. **(b)** On the average of 4 times a day the land somewhere in Japan rocks and rolls with an earthquake. The 4 main Japanese islands are part of a chain of rather recently formed volcanic mountains, and much of the land is covered with volcanic ash and lava.

3. **(a)** The Mariana Trench is the deepest place in the oceans of the world: 36,198 feet, extending from southeast of Guam to northwest of the Mariana Islands, in the western Pacific.

4. **(c)** Ice covers one-tenth of the land surface of the planet.

5. **(d)** The North Atlantic island of Iceland has more than 100 volcanoes and about 120 glaciers. The volcano Hekla has had 21 eruptions in the last 8 centuries. Only a seventh of Iceland is agriculturally productive.

6. **(d)** The 260-mile-long Beardmore Glacier, the world's largest glacier, is in the Queen Maud Mountains of Antarctica.

7. **(d)** The active, 12,450-foot volcano Mount Erebus is on James Ross Island in Ross Sea, Antarctica.

8. **(a)** At the peak of the most recent Ice Age, which began about 2 million years ago, the sea level was perhaps 330 feet lower than it is today, because tremendous amounts of water were frozen in the glaciers that covered much of the land. When the Earth warmed again and the Ice Age began retreating, 18,000 years ago, the sea level rose as the glaciers melted.

9. **(b)** There are 32 seas on the Earth, in addition to the 4 oceans.

10. **(c)** The "Antarctic Donut" is a huge hole in the ozone over the South Pole that is believed to be caused by the release of chlorofluorocarbons from aerosol products.

11. **(d)** Lascar, in the Andes Mountains in central Chile, is the highest active volcano in the world: 18,077 feet.

12. **(d)** Mount Saint Helens, in the state of Washington, had been dormant since 1857 when its eruption in 1980 sent plumes of steam more than a mile high and dropped ash 50 miles away.

13. **(b)** Ocean basins, the largest depressions in the Earth, are created because the rock that makes up the continents is lighter and higher than the basaltic rock that underlies the oceans.

14. **(a)** The lithosphere is the solid outer part of our planet. The atmosphere is the layer of air that extends above the lithosphere. The Earth's water—on the surface, in the ground, in the air—makes up the hydrosphere. The biosphere is about 12 miles from top to bottom. Most life exists on and in the lithosphere and in 400 feet of the hydrosphere.

15. **(a)** The Ring of Fire is a belt of hundreds of active volcanoes from the southern tip of South America north to Alaska, then west to Asia, and south through Japan, the Philippines, Indonesia, and New Zealand. It marks the boundary where the plates that cradle the Pacific Ocean meet the plates that hold the continents surrounding the ocean.

16. **(d)** Prime Meridian, or the Meridian of Greenwich, is in London's borough of Greenwich.

17. **(c)** Australia's 4 great desert regions are the Great Sandy, the Outback, the Gibson, and the Great Victoria.

18. **(b)** The desert along the southwest coast of Africa is noted for the mining of the precious gem diamonds.

19. **(d)** There are about 600 active volcanoes on the face of the Earth.

20. **(d)** About 70 percent of the Earth's surface is covered with water.

21. **(b)** The Equator, the imaginary circle that divides the planet into northern and southern hemispheres, is exactly 24,901.5 miles in length.

22. **(d)** It was in 1976 that the Concorde began cutting flight time across the Atlantic Ocean in half.

23. **(a)** The deepest chasm in North America is Hells Canyon, 7,900 feet deep, on the Idaho-Oregon border. The Grand Canyon, in Arizona, is "only" about a mile deep.

24. **(d)** A desert is an area of land that receives less than 10 inches of precipitation a year.

25. **(c)** The Himalayas, the world's highest mountains, owe their origin to the collision of the Eurasian and Indo-Australian tectonic plates.

26. **(a)** Israel and Jordan share the Dead Sea, whose surface is 1,302 feet below the level of the nearby Mediterranean Sea.

27. **(d)** The greatest known depth of the Indian Ocean is nearly 5 miles: 25,344 feet.

28. **(d)** Between 69 and 70 degrees separate the parallel lines of latitude. It is also about 70 miles between degrees of longitude at the Equator, but the distance between lines of longitude shrinks to just a foot or two near the North Pole and the South Pole.

29. **(d)** "Vog," for volcanic fog, describes the soupy hazes from the 1,500 to 2,000 metric tons of volcanic gases that pour into the Hawaiian air from the Kilauea volcano. Vog has at times spread northwest across the island chain as far as Honolulu, 150 miles away, and caused respiratory problems for the elderly and the young. Volcanic gases produce acid rain nearly as concentrated as that in the industrial east coast of the United States.

30. **(c)** Bering Glacier, the largest glacier in North America, is 126 miles long and about 30 miles wide near its terminus, in the Chugach-Saint Elias Mountains in southern Alaska.

31. **(b)** Nicaragua's unique gap, the longest lowest gap in the western hemisphere, prompted engineers to consider the site as the western part of a canal between the Atlantic and the Pacific Oceans. Frequent volcanic eruptions in the area foreclosed the possibility.

32. **(a)** Loess is the name for fine, silty sediment deposited by wind. In northern China, where it covers a vast area, the loess is a fine loam, rich in lime and yellowish in color.

33. **(b)** Citlaltepetl, Mexico's highest point, is an 18,700-foot-high volcanic peak in central Veracruz.

34. **(d)** Mount Etna's 140 eruptions have included notably destructive ones in the years 1169, 1669, and 1852.

35. **(d)** Popocatepetl, which last erupted in 1702, and Iztaccihuatl, which last erupted in 1868, are side by side in the central region of Mexico. Popocatepetl is only 2,433 feet lower than North America's tallest mountain, 20,320-foot-high Mount McKinley, in Alaska.

- One of the potentially richest countries in sub-Saharan Africa is Angola. It has extensive petroleum potential, rich agricultural land, and valuable mineral resources. Its economy was disrupted severely when most of the 350,000 Portuguese who had run the nation departed as Angola gained its independence.
- Only 30 out of every 100 Europeans are under 20 years of age. More than 50 percent of all Africans are less than 20.
- A nation that has become independent within the last few years is likely to be an island or an archipelago.
- Antarctica's largest land animal is a wingless insect less than one-tenth inch long.
- Two million Bengalis starved to death when the Japanese military cut off the flow of rice from Burma to Bengal in 1943, during the Second World War.
- Virtually all of Maine is north of Cape Sable, Nova Scotia.
- Southernmost Canada is south of northernmost Pennsylvania. It is within 138 miles of the Mason-Dixon line.
- The entire state of Connecticut and Cape Cod are both south of northernmost Pennsylvania.
- Due south of Lake Superior is Rome, Georgia.
- All of Chesapeake Bay is north of Cairo, Illinois.
- The western tip of Virginia is 25 miles west of Detroit, Michigan.
- Virginia's northern tip is north of Atlantic City, New Jersey.
- Atlanta, Georgia, is closer to Detroit, Michigan, and to Chicago, Illinois, than it is to Miami, Florida.
- Arranging the world's population in a crowd with about 1 square yard of space per person would fill Long Island, New York.
- Most of the rain clouds that form over the Pacific Ocean are prevented by the Cascades and Sierra Nevada mountains from reaching the Great Basin, which lies between the Pacific mountain ranges and the Rocky Mountains in the western United States. The Mojave Desert, in southern California, is the dry southern end of the Great Basin.

- Tens of thousands of years ago, the Sahara regions were covered with green vegetation and filled with game.
- Life expectancy in the Islamic republic of Mauritania, part of the vast western Saharan shield of crystalline rocks, is 46 years. The northern four fifths of the country is barren desert.
- All Asia lies to the west of the International Date Line.
- The People's Republic of China's 14,000-mile border is shared by Vietnam, Laos, Myanmar, Bhutan, Nepal, India, Pakistan, Korea, the USSR, Mongolia, and Afghanistan.
- At Climax, in central Colorado, is the world's largest molybdenum mine. Molybdenum is a hard, silvery, metallic element used as an alloy in strengthening and hardening steel.
- Less than a quarter of Hawaii's original forests remains, and about 40 percent of the state's native bird species have been driven into extinction.
- Ninety percent of El Salvador is of volcanic origin.
- Yakutsk is the only large town on the 2,652-mile-long Lena River in the east-central Soviet Union.
- Bolivia lost its Pacific seacoast in the 1879-84 War of the Pacific with Chile.
- Costa Ricans are overwhelmingly of European descent; Spain is the primary country of origin.
- Since the early 1980s, Botswana has become the world's largest producer of quality diamonds.
- The city of Manus is 1,000 miles from the mouth of the Amazon River and accessible to ocean steamers.
- Manx, now virtually extinct, was the language of the Isle of Man in the Irish Sea off the northwest coast of England.
- Kalaupapa is a leper settlement on the northern coast of Molokai Island in the Hawaiians.
- Not known until the 1900s were the Maoke Mountains in Indonesia. They run east and west from the boundary of Papua New Guinea; the highest point is Djaja, 16,535 feet.

- No official census has been taken in Lebanon since 1932.
- The Amazon River was originally called the Orellana. Perpetual free navigation of the river has been guaranteed by treaty between Colombia and Brazil.
- The coasts of the Leeward Island Saba, in the Netherlands Antilles, are sheer cliffs about 2,800 feet high.
- The term "new world" to describe the western hemisphere and the discovery of America was first used in 1516 by the Italian historian Peter Martyr.
- Because it followed a pro-Axis policy during the Second World War, Spain was not allowed to join the United Nations until 1955.
- The People's Republic of China has the world's largest army: some 3 million under arms, some 5 million in the reserves.
- Dust, ashes, and smoke rose to a height of about 17 miles during the most violent eruption of modern times—Krakatau's, in 1883—36,000 people in Indonesia were killed, and a tidal wave about 50 feet high was formed.
- Earth once had a 435-day year.
- Nearly three quarters of Spain is arid.
- Greenland's Cape York is noted for its large iron meteorites.
- Yellowstone Lake, at 137 square miles, is the largest body of water in North America at so great an altitude: 7,735 feet.
- Eighteen thousand people were killed when an avalanche of rocks and ice rolled down the Andes in 1970 and overwhelmed the Peruvian town of Yungay.
- The Amazon River basin drains an area roughly three fourths the size of the contiguous United States.
- La Mancha, the south-central Spanish region made famous by Cervantes, is a high, level, arid, treeless plateau.
- The largest lake in Europe is Ladoga, 6,835 square miles, in the western Soviet Union.
- The state of Missouri's motto is "Let the Welfare of the People Be the Supreme Law."

- The median temperature of the planet is about 60° F.
- Two million years ago, Earth had the same volume of water as it has today.
- The planet's oldest ecosystems are coral reefs, complex associations of plants and animals; some were formed nearly 600 million years ago.
- The Himalaya Range is still building.
- A future Hawaiian island, Loihi, has risen to 3,000 feet beneath the surface of the Pacific Ocean south of Hawaii.
- Fourteen peaks of the Himalaya Range are over 25,000 feet.
- Africa's Lake Victoria, the main reservoir of the Nile River, is almost as big as Scotland.
- Literally in the middle of nowhere, Ayers Rock towers 1,100 feet above flat central Australia, in the southwest Northern Territory.
- The African elephant is the largest of all land animals.
- Cuba and Puerto Rico, the last of Spain's colonies in the western hemisphere, became US dependencies at the very end of the nineteenth century.
- The area around the North Pole is entirely water, usually ice-covered.
- Gudena, the longest river in Denmark, is 98 miles long.
- The volcano Paricutín, 200 miles west of Mexico City, grew to a height of 1,500 feet above its base 8 months after forming.
- More than 90 percent of the people of Nepal are engaged in agrarian pursuits.
- The highest peak in Europe outside of the Alps and the Caucasus is Mulhacén, in the Sierra Nevada, in southern Spain: 11,407 feet.
- The largest city north of the Arctic Circle is the Soviet Union's Murmansk, an ice-free port about 22 miles from the Arctic Ocean; it was a major supply center for Allied convoys during the Second World War.
- About 3 times the size of California, Africa's Niger is about two-thirds desert and mountains, one-third savanna.
- Ethiopia, which is about the size of Texas, Oklahoma, and New Mexico combined, is the oldest independent country in Africa, and one of the oldest in the world.
- Two thirds of Yugoslavia is mountainous.

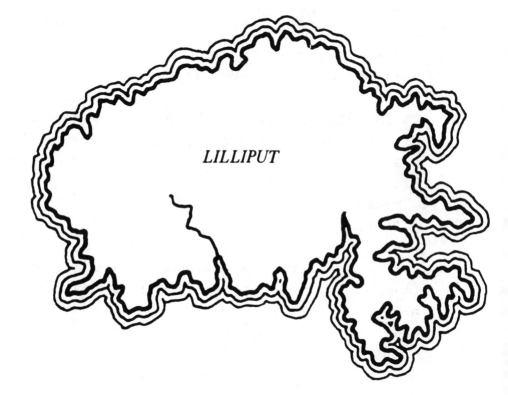

LILLIPUT

1. The oldest European settlement in the Far East is
    a. Macao
    b. Hong Kong
    c. Taiwan
    d. Luzon

2. The 2 republics that share the island of Hispaniola are
    a. Saint Lucia and Grenada
    b. Antigua and Barbuda
    c. Trinidad and Tobago
    d. Haiti and the Dominican Republic

3. The South Pacific island of Nauru has one of the highest per-capita incomes in the world because of the _____ mined there.
    a. gold
    b. phosphate
    c. nitrate
    d. uranium

4. The Soviet Union's Big Diomede Island and the United States' Little Diomede Island are _____ miles apart in the Bering Sea.
    a. 2
    b. 19
    c. 40
    d. 312

5. This atoll in the middle of the Pacific Ocean was designated by the US Atomic Energy Commission in 1947 as a test site for atomic weapons:
    a. Eniwetok
    b. Wake
    c. Guam
    d. Bikini

6. The largest island in the Mediterranean Sea is
   a. Crete
   b. Corsica
   c. Sardinia
   d. Sicily

7. New Guinea, the second largest island in the world, and the Fiji Islands are part of
   a. Melanesia
   b. Micronesia
   c. Polynesia
   d. the Channel Islands

8. The Hebrides are islands near
   a. India
   b. Scotland
   c. Argentina
   d. Malaysia

9. Because of its strategic importance, the most bombed place in the Second World War was
   a. Iceland
   b. Malta
   c. Sardinia
   d. Okinawa

10. Kalatdilt Nunat is the native name for
    a. Borneo
    b. Japan
    c. Greenland
    d. Mozambique

11. The Moluccas, islands in Indonesia, are also known as
     a. Eldorado
     b. God's Revenge
     c. the Spice Islands
     d. the Isles of Gold

12. The Dodecanese are
     a. islands south of Australia
     b. islands in the Aegean Sea
     c. islands that sank in the Atlantic Ocean
     d. the geographic term for mountainous islands

13. There are 700 Bahama islands. _____ are inhabited.
     a. All 700
     b. Half of them
     c. All but 7
     d. Only 29

14. _____ is geographically the largest and also the most populous democracy in the Caribbean.
     a. Barbados
     b. Jamaica
     c. Haiti
     d. The Dominican Republic

15. Most of Japan's population lives on 1 of the country's 4 main, mountainous islands:
     a. Honshu
     b. Shikoku
     c. Kyushu
     d. Hokkaido

16. Greenland, the largest island in the world, is 839,999 square miles. About _____ square miles lie under a permanent icecap.
   a. 25,000
   b. 100,000
   c. 700,000
   d. 820,000

17. Indonesia, formerly the Dutch East Indies, is an archipelago in southeast Asia. The republic consists of _____ islands.
   a. exactly 50
   b. exactly 100
   c. about 1,000
   d. 13,667

18. Halmahera, about 6,928 square miles, is the largest island of the _____ group.
   a. Caroline
   b. Mariana
   c. Gilbert
   d. Molucca

19. Argentines refer to the group of islands some 490 miles northeast of Cape Horn, at the southern tip of South America, as the Malvinas. The British call them the
   a. Maldives
   b. Perons
   c. Falklands
   d. Wellingtons

20. Viti Levu is the largest island in this group in the southwest Pacific Ocean:
   a. Gilbert   **c.** Fiji
   b. Marshall   **d.** Caroline

21. The world financial center of Singapore, which is on an island called Singapore in the independent country of Singapore, is located off the southern tip of this peninsula:
    a. Malay
    b. Gaspé
    c. Cape York
    d. al Cape Hadd

22. Brunei is a small, oil-rich country on the northern coast of the third largest island in the world:
    a. Borneo
    b. Hispaniola
    c. Sri Lanka
    d. Celebes

23. _____ is known as the "island of stone money."
    a. Easter Island
    b. Yap Island
    c. Christmas Island
    d. Wake Island

24. North Island and South Island are the principal islands of
    a. Indonesia
    b. the Falklands
    c. Singapore
    d. New Zealand

25. The Philippines in the western Pacific Ocean are an archipelago of about _____ islands lying approximately 500 miles off the southeast coast of Asia.
    a. 100
    b. 750
    c. 5,000
    d. 7,100

26. Oceania, not including Australia and New Zealand, consists of about
    _____ islands scattered throughout the Pacific Ocean with a
    population of 5 million.
    a. 2,000
    b. 7,500
    c. 13,000
    d. 20,000

27. Two Caribbean islands are overseas departments of France:
    a. Guadeloupe and Martinique
    b. Haiti and the Dominican Republic
    c. Saint Vincent and the Grenadines
    d. Trinidad and Tobago

28. Most of Polynesia is east of the International Date Line and
    includes Easter Island, Samoa, Tahiti, and
    a. Hawaii
    b. New Guinea
    c. the Fiji Islands
    d. Okinawa

29. Micronesia is in
    a. the Pacific Ocean
    b. the North Atlantic Ocean
    c. the Mediterranean Sea
    d. the central Indian Ocean, south of India

30. The largest island in Europe is
    a. Iceland
    b. Great Britain
    c. Malta
    d. Shetland

31. In the Niagara River, in western New York state, this island divides Niagara Falls into the American Fall and the Horseshoe Fall:

a. Niagara Island

b. Goat Island

c. Toronto Island

d. Maple Leaf Island

32. The second largest island in the world, New Guinea, and the tropical jungles of Cape York Peninsula in northeast Australia are only _____ miles apart.

a. 2

b. 7

c. 13

d. 100

1. **(a)** Macao, about 40 miles west of Hong Kong off the coast of the People's Republic of China, became a Portuguese trading post in 1557, making the 6-square-mile island the oldest European settlement in the Far East. It relies today on China for its drinking water and for much of its food supply.

2. **(d)** Haiti and the Dominican Republic are the 2 republics that share the Caribbean island of Hispaniola.

3. **(b)** The mining of phosphate, derived from decayed marine organisms and bird droppings, has made Nauru a very rich country per capita.

4. **(a)** The Diomedes, which are only 2 miles apart, are separated by the International Date Line. They were discovered and named on St. Diomede's Day, 1728, by the Danish explorer Vitus Bering in the employ of Russia.

5. **(a)** Eniwetok, since 1947 a US proving ground for atomic weapons, is an atoll at the northwest end of the Ralik chain in the northwest Marshall Islands. Circular in shape, Eniwetok has 40 islets.

6. **(d)** Sicily (9,822 square miles) is the largest island in the Mediterranean Sea.

7. **(a)** The Melanesia island group in the southwest Pacific Ocean includes New Guinea and the Fiji Islands.

8. **(b)** The Hebrides, or Western Islands, 2,900 square miles, are in the Atlantic Ocean west of Scotland.

9. **(b)** The 3 islands of Malta, just south of Sicily, in the Mediterranean Sea, were bombed by Axis planes about 1,200 times during the Second World War. Because the islands were a strategic British base between southern Europe and North Africa, Malta was the most bombed place in the war. US President Franklin D. Roosevelt hailed Malta as "one tiny bright flame in the darkness."

10. **(c)** Kalatdilt Nunat is the native name for Greenland.

11. **(c)** Native to the Moluccas, or Spice Islands, is the only tree in the world that bears a seed from which 2 spices are extracted: nutmeg and mace come from the *Myristica fragrans.*

12. **(b)** A dozen main islands and numerous small islands, the Dodecanese are in the southeast Aegean Sea and are a department of Greece.

13. **(d)** Only 29 of the 700 islands of the Bahamas are inhabited; the island chain in the western Atlantic Ocean also has more than 2,000 cays and dry rocks.

14. **(d)** The Dominican Republic—18,704 square miles, about 6.5 million people—the eastern two thirds of the island of Hispaniola—is geographically the largest and also the most populous democracy in the Caribbean.

15. **(a)** Honshu is far and away the most populous of the 4 main islands of Japan.

16. **(c)** About 700,000 of Greenland's 839,999 square miles lie under a permanent icecap.

17. **(d)** About 153 million people are scattered over Indonesia's 13,667 islands.

18. **(d)** Halmahera, the largest island of the Moluccas, in Indonesia, lies on the Equator.

19. **(c)** The South Atlantic islands the British call the Falklands are called the Malvinas by the Argentines.

20. **(c)** Viti Levu is the largest island in the Fiji group in the southwest Pacific Ocean.

21. **(a)** Singapore is off the southern tip of the Malay Peninsula, in southeast Asia.

22. **(a)** Oil was discovered in Brunei, on the northern coast of Borneo, in 1929. (Sharing the northern part of the world's third largest island, 290,320 square miles, are the Malaysian states of Sabah and Sarawak.)

23. **(b)** Yap, the largest of the tiny Yap Island group in the western Pacific Ocean, is notable for large pieces of circular stone money, remains of an earlier time.

24. **(d)** North Island and South Island are the 2 principal islands of New Zealand. The capital, Wellington, is on North Island, whose other chief cities are Auckland, Manukau, and Hamilton. South Island's chief cities are Christchurch, Dunedin, Nelson, and Timaru.

25. **(d)** The major islands of the Philippines' 7,100 are Cebu, Leyte, Luzon, Mindanao, Panay, and Negros.

26. **(d)** Oceania consists of about 20,000 islands in the Pacific Ocean.

27. **(a)** The 2 Caribbean islands that are overseas departments of France are Guadeloupe and Martinique.

28. **(a)** Hawaii, Samoa, Easter Island, and Tahiti are part of Polynesia, a subdivision of Oceania in the central Pacific Ocean, between 30° North and 47° South latitude.

29. **(a)** Micronesia is in the central Pacific Ocean and includes the Marshall, Caroline, and Gilbert Islands.

30. **(b)** Great Britain, comprising England, Scotland, and Wales (together with Northern Ireland they constitute the United Kingdom of Great Britain and Northern Ireland), is the largest island in Europe: 93,598 square miles, with a population of 56 million.

31. **(b)** Goat Island, which is three quarters of a mile long in the Niagara River just above Niagara Falls, on the New York-Canadian border, divides the falls into American Fall and Horseshoe Fall.

32. **(d)** Only 100 miles—Torres Strait—separate New Guinea, the second largest island in the world, from northeast Australia.

- The Soviet Union's national district Taimyr National Okrug lies entirely within the Arctic Circle and has generous deposits of gold, nickel, platinum, and copper. The population is less than 40,000.
- Mount Pelée's eruptions in 1902 killed more than 30,000 people on the West Indies island of Martinique and destroyed the city of Saint Pierre.
- The oldest rock found on the ocean floor dates back only about 150 million years. Rock in northern Canada dates back 4 billion years.
- The world's leading producer of rubber, tin, palm oil, and tropical timber is Malaysia.
- Communism Peak is the highest mountain in the Soviet Union. Lenin Peak, formerly Kaufmann Peak, is the second highest.
- Coal now being formed in taigas and swamps will not be ready for use for millennia.
- The southern Sahara's largest lake, Chad, was once as big as the state of New Jersey and a major source of sustenance for one of Africa's poorest regions. Lake Chad has been shrinking into extinction since 1970.
- In 1983 a forest fire in Borneo consumed an area the size of the state of Connecticut.
- The eruption of Indonesia's Mount Tambora in 1816 led to the "year without summer" in the northeastern United States and northern Europe. Global temperature dropped an estimated 3.5° F.
- Because the Soviet Union continues to occupy the Northern Territories—small islands off the coast of Japan's Hokkaido island—relations between Moscow and Tokyo have not been close.
- Saint Helena, the British island in the South Atlantic where Napoleon spent the years 1815–21 in exile, was a detention camp for Boer prisoners in the period 1899–1902.
- Niagara Falls was created by a glacier.
- It is said that Darjeeling, a town in northeast India, commands one of the finest views in the world; it includes Mount Everest, visible on clear days 110 miles to the northwest.

MQ20. The diagonally marked area represents the western half
of the second largest island in the world:

a. Irian Barat

b. Haiti

c. Ireland

d. Labrador

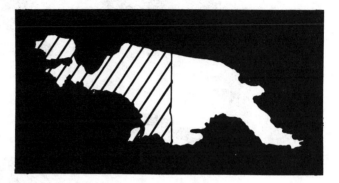

MQ21. This is a map of the island chain of

a. Indonesia

b. Japan

c. the Aleutians

d. the Maldives

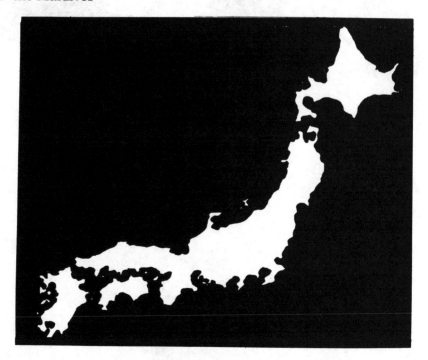

MQ22. Match the 5 islands listed:

Cuba	1. _____
Hokkaido	2. _____
Iceland	3. _____
Trinidad	4. _____
Newfoundland	5. _____

MQ23. Match the 5 islands listed:

Sri Lanka   1. _____

Sulawesi (Celebes)   2. _____

Sicily   3. _____

Sardinia   4. _____

Malagasy Republic (Madagascar)   5. _____

MQ20. **(a)** Diagonally marked Irian Barat, or West Irian, or West New Guinea, formerly Dutch New Guinea, is a 162,927-square-mile province of Indonesia. It is the western half of New Guinea, the second largest island in the world. The independent country of Papua New Guinea occupies the eastern half.

MQ21. **(b)** Japan's mountainous, volcanic 4 main islands, in the western Pacific Ocean, are from north to south Hokkaido, Honshu, Kyusu, and Shikoku: 143,619 square miles, with a population of more than 120 million. Japan, which is slightly smaller in area than California, has a density of about 844 people per square mile.

MQ22. The 5 islands are: 1. Hokkaido, the northernmost of the 4 main islands of Japan in the Pacific Ocean off the east coast of Asia; 2. Newfoundland, a province of Canada in the Atlantic Ocean; 3. Cuba, 746 miles long in the Greater Antilles, West Indies, south of Florida and north of the western Caribbean Sea. (The US naval station at Guantanamo Bay is on the island's southeast coast.); 4. Trinidad, in the West Indies off the northeast coast of Venezuela, was discovered by Columbus in 1498; 5. Iceland, between the North Atlantic and Arctic Oceans, has suffered from destructive earthquakes; it has 120 glaciers and more than 100 volcanoes, and only a seventh of the land is productive.

MQ23. The 5 islands are: 1. Sardinia, a 9,301-square-mile island in the Mediterranean Sea west of the southern Italian peninsula. During the Second World War it was a German air base; 2. Sri Lanka, which is about the size of West Virginia, is an Indian Ocean island-state southeast of India; the 2 countries are connected by a chain of shoals called Adam's Bridge. The 25,332-square-mile island, known as Ceylon before 1972, is rich in tropical vegetation. Tamil separatists and government forces have warred violently since 1983; 3. Sicily is the largest island in the Mediterranean Sea: 9,925 square miles. It is west of the extreme southern point of the Italian peninsula; 4. Sulawesi is the Indonesian name for the island of Celebes in the Malay archipelago east of Borneo: 69,255 square miles. Chief products include cassava, yam, sulfur, and corn; 5. Malagasy Republic, or Madagascar, slightly smaller than Texas at 226,657 square miles, is the fourth largest island in the world. It is separated from the southeast African coast by the Mozambique Channel. Plateaus and mountains are topographical features. Chief products: soap, mica, cement.

Embarrass is an 185-mile-long river in eastern Illinois. It flows south and southeast into the Wabash River.

Goodnews Bay is a village on an inlet of the Bering Sea in southwest Alaska.

Indefatigable is an island in the Galápagos, in the eastern Pacific Ocean.

Tandjungpandan is the chief town of Belitung, a 55-mile-long Indonesian island in the Java Sea off the southeast coast of Sumatra.

Rabbit Ears Pass is a mountain highway pass, at 9,572 feet, in northern Colorado.

Great Slave Lake, 10,980 square miles, is in the Northwest Territories, Canada.

Hackney is a borough of Greater London, England.

Ompompanoosuc is a small river in eastern Vermont; it enters the Connecticut River about 17 miles north of White River Junction.

Prettyboy Reservoir is the chief reserve for the water supply of Baltimore, Maryland.

The **Bug** is a 481-mile-long river in east-central Poland.

Treasure is a county in Montana.

Winnibigoshish is a 14-mile-long lake in north-central Minnesota.

The **Danger Islands** are an island group in the Northern Cook Islands, in the central Pacific Ocean.

Going-to-the-Sun Mountain is a 9,615-foot peak in Glacier National Park, in northwest Montana.

Geographe Bay is an inlet of the Indian Ocean in southwestern Western Australia.

Nonsuch is a small island in Bermuda.

No is a lake in south-central Sudan, in east Africa.

Mount Mother is a volcano at the northeast tip of New Britain Island in the Bismarck Archipelago.

Aa is a small river in northern France; it flows into the North Sea.

Phra Nakhon Si Ayutthaya is a city on an island in the lower Chao Phraya River, in southern Thailand.

Mount Despair is an 8,585-foot-high peak in Glacier National Park, in northwest Montana.

Knockadoon Head is a cape on the southern coast of Eire.

Clouds Rest is a 9,930-foot-high mountain in central California.

Thayawthadangyi is an island in a Burmese (Myanmarese) archipelago.

Thousand Islands is a group of about 1,500 islands in the upper Saint Lawrence River, in Canada and the state of New York.

Pangkalanberandan is a town with large oil fields in northern Sumatra, Indonesia, near the northern end of the Strait of Malacca.

Cannonball is a 140-mile-long river in southwest North Dakota.

Cape Catastrophe is at the western entrance to Spencer Gulf, in South Australia.

The Father is an active volcano in the Whiteman Range on the island of New Britain in the western Pacific.

Memphremagog is a 30-mile-long lake extending from northern Vermont into southern Quebec; about 23 miles are in Canada.

Paw Paw is a village in southwest Michigan.

Male is the chief atoll and capital of the Maldives, 19 clusters of coral atolls in the Indian Ocean.

Arctic Red is a 310-mile-long river in Canada's Northwest Territories.

Arafat is a granite hill 15 miles southeast of Mecca, Saudi Arabia, a goal of pilgrimages.

Villa Hayes, named for US President Rutherford B. Hayes, who arbitrated an Argentine-Paraguayan boundary dispute, is a town on the Paraguay River in west-central Paraguay.

Sweet Home is a city of 7,000 in western Oregon.

Opportunity is an urban community east of Spokane, Washington.

Flaming Cliffs is the highland in Mongolia where fossils and dinosaur eggs were discovered in 1925.

Seydhisfjördhur is a coastal town in Iceland.

Thousand Ships Bay is in the Solomon Islands, in the western Pacific Ocean.

Thickanetley Bald is a 4,054-foot-high peak in northern Georgia, in the southeast United States.

Sound of Rum is a channel between the Isle of Rum and Eigg Island in the Inner Hebrides, off the west coast of Scotland.

Lake Pun Run is at an altitude of about 14,200 feet in central Peru.

Deep Bottom is a hamlet in east-central Virginia.

Potato Knob is a 6,420-foot-high mountain in western North Carolina.

Cockermouth, the birthplace of the English poet William Wordsworth, is an urban district in northwest England.

Kirkcudbrightshire was John Paul Jones's home town in Scotland.

Devil's Ear Mountain is a 3,903-foot-high peak in the Adirondack Mountains, in northeast New York.

Social Circle is a small city in north-central Georgia, 38 miles east-southeast of Atlanta.

Widows' Tears is an 1,170-foot waterfall in Yosemite National Park, in east-central California.

Standing Indian is a 5,500-foot-high mountain in southwest North Carolina.

The **Friendly Islands** are a 270-square-mile archipelago of about 150 islands in the southwest Pacific Ocean; they are also called the Tonga Islands.

The **Withlacoochee** is a 120-mile-long river in western Florida.

Soddy Daisy is a small city in southeast Tennessee, 16 miles northeast of Chattanooga.

Sombrero is a small island, a part of Saint Christopher-Nevis, in the West Indies. (Its reserves of phosphate have been mined out.)

Smiley Mountain, in central Idaho, is 11,506 feet high.

Tongue is a 246-mile-long river in southeast Montana.

Grapevine is a city 20 miles northeast of Fort Worth, in northern Texas.

Happy Valley is a city of 4,000 in Newfoundland, in eastern Canada.

Ragged Island, about 5 square miles, with a population of about 400, is one of the Bahama Islands, north of the east end of Cuba, in the Atlantic Ocean.

The **Porcupine Mountains,** with their highest peak at 1,958 feet, are in the northwest extremity of the upper Michigan peninsula.

No Mans Land is a small island southwest of Martha's Vineyard in the Atlantic Ocean.

Rosebud is a 100-mile-long creek in southeast Montana.

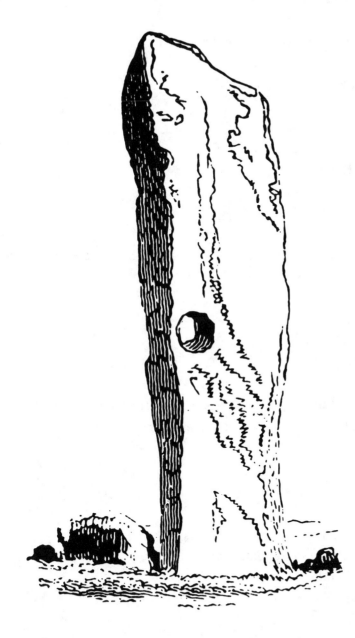

1.   Boothia Peninsula is
     a. the northernmost mainland point of North America
     b. the dividing site between Iran and Iraq
     c. the ancients' name for Italy
     d. the largest landmass in the Philippines.

2.   Denmark Strait separates _____ and _____.
     a. Australia . . . New Guinea
     b. Japan . . . Korea
     c. Greenland . . . Iceland
     d. Sweden . . . Finland

3.   The narrow landform with water on both sides that connects
     2 larger land areas is called
     a. an atoll
     b. an isthmus
     c. an islet
     d. a cay

4.   Between the northern tip of Oman, on the southeastern Arabian
     Peninsula, and the southern coast of Iran is the Strait of
     a. Oman
     b. Emir
     c. Hormuz
     d. Persia

5.   Africa and the Arabian Peninsula are divided by
     a. the Persian Gulf
     b. the Red Sea
     c. the Suez Canal
     d. the Nile River

6. The Canadian province with the largest land area is
    a. Quebec
    b. the Northwest Territories
    c. Manitoba
    d. Saskatchewan

7. The largest deposit of coral in the world is
    a. around the Canary Islands
    b. in Hudson Bay, Canada
    c. off the southeast coast of Africa
    d. off the east coast of Australia

8. Hoba West is
    a. the largest gold mine in the world
    b. the longest river in Borneo
    c. the largest known meteorite
    d. the most eruptive of all volcanoes

9. The Lascaux cave complex with prehistoric paintings is in
    a. Nigeria
    b. France
    c. Thailand
    d. Morocco

10. The Gulf of Bothnia is bordered by
    a. Egypt and Libya
    b. Panama and Colombia
    c. Sweden and Finland
    d. Mozambique and Madagascar

11. Manhattan, Singapore, Lagos (the capital of Nigeria), and Montreal (Canada's largest city) share a geographic feature. They
    a. have each been host to a world's fair
    b. are on islands
    c. have more Chinese people than black people
    d. have each exactly 500 miles of paved roads

12. The Panama Canal eliminates _____ miles from the sea voyage between New York and San Francisco.
    a. 7,800
    b. 10,008
    c. 12,750
    d. 15,225

13. The Strait of Gibraltar, the 36-mile-long waterway that links the Atlantic Ocean and the Mediterranean Sea, is 23 miles at its widest. It is about _____ miles wide at its narrowest part.
    a. 2
    b. 4
    c. 8
    d. 19

14. "Ria" is a geographical term meaning
    a. a drowned river
    b. tectonic activity
    c. lake evaporation
    d. valley

15. The Rock of Gibraltar, the mountain on the east end of the Strait of Gibraltar, which is part of the British naval and air base of Gibraltar, is _____ feet high.
    a. exactly 100   c. 1,398
    b. 349   d. exactly 3,000

16. It wasn't until the year _____ that South Africa's Cape of Good Hope was first passed by a European sailor.
    a. AD 853
    b. 1241
    c. 1497
    d. 1603

17. Pomerania is a historical region on the
    a. east coast of Argentina
    b. Adriatic coast of Yugoslavia
    c. Baltic Sea
    d. northern shore of the Great Lakes

18. In northern Africa and the Near East, the Arabic term *wadi* in place names means
    a. oasis
    b. plateau
    c. valley
    d. mountain peak

19. "Disappointment" is the name of a cape
    a. in the state of Washington
    b. in northwestern Scotland
    c. in the Aleutian Islands
    d. at the southern tip of India

20. The world's largest contiguous plantations are in
    a. Brazil
    b. Indonesia
    c. Zambia
    d. Liberia

21. The world's highest waterfall is in
    a. Venezuela
    b. Bali
    c. Hungary
    d. Canada

22. The Kamchatka Peninsula separates the Sea of Okhotsk from the
    a. Red Sea
    b. Bay of Biscayne
    c. Aegean Sea
    d. Bering Sea

23. The longest undefended border in the world is the frontier between
    a. the Soviet Union and the People's Republic of China
    b. Brazil and Bolivia
    c. the United States and Canada
    d. India and the People's Republic of China

24. This moist, hot lowland bordering the Atlantic Ocean was long known as the Slave Coast:
    a. western Africa along the Bight of Benin
    b. South Africa
    c. Morocco
    d. Spanish Sahara

25. The Seikan rail tunnel beneath Tsugaru Strait connecting cold, northern Hokkaido island and Japan's main island of Honshu is
    a. 5.78 miles long
    b. 9 miles long
    c. 10.9 miles long
    d. about 33 miles long

26. Chuquicamata, in northern Chile, is
   a. the earthquake capital of the world
   b. the volcano capital of the world
   c. the largest known single copper-mining property in the world
   d. a nuclear-free zone

27. Ethiopia's access to the Red Sea is through
   a. Eritrea
   b. Somalia
   c. Tanzania
   d. Mozambique

28. The overland trade routes that linked the Mediterranean world with China many centuries ago were called
   a. the Lido Road
   b. the Avenue of the Polos
   c. the Silk Route
   d. the Emerald Highway

29. The eastern terminus of the Trans-Siberian Rail Road and the principal Soviet seaport on the Pacific Ocean is
   a. Vologda
   b. Vladivostok
   c. Vorkuta
   d. Voroshilovgrad

30. The narrow strait that separates Turkey in Asia from Turkey in Europe is the
   a. Bosporus
   b. Marmara
   c. Constantinople
   d. Socrates

31. Recife is the easternmost point of
    a. North America
    b. South America
    c. Asia
    d. Antarctica

32. This westernmost Canary Island was thought by ancient geographers to be the western limit of the world:
    a. Tenerife
    b. Hierro
    c. La Palma
    d. Gomera

33. The low-level treeless plains known as the Barren Grounds are in
    a. Chile
    b. Australia
    c. New Zealand
    d. Canada

34. This seaport handles more cargo than any other in the world:
    a. Los Angeles
    b. Tokyo
    c. Canton
    d. Rotterdam

35. Cape Horn is at the southern extremity of
    a. Africa
    b. Panama
    c. South America
    d. Sri Lanka

36. The Arabian Gulf, an arm of the Arabian Sea, is also known as

    a. the Red Sea

    b. the Suez Canal

    c. the Persian Gulf

    d. the Sea of the Emirs

37. The swampy region in Central America known as the "Mosquito Coast" is in

    a. Panama

    b. Honduras

    c. Guatemala

    d. Nicaragua

38. The Barbary States were

    a. California, Oregon, and Washington

    b. Morocco, Algiers, Tunis, and Tripoli

    c. French Guiana, Dutch Guiana, and Guyana

    d. Virginia, North Carolina, and South Carolina

39. Infamous Transylvania is a region in

    a. Siberia

    b. Australia

    c. Romania

    d. the Himalayas

40. The Hitler Line was a

    a. German defensive fortification in western Italy in the Second World War

    b. German defensive fortification on the Atlantic seaboard, in Belgium

    c. moat around Berlin

    d. name for the Maginot Line in Nazi hands

41. White Man's Grave is a name formerly applied to the lands along

    a. the southeast coast of the United States

    b. the Guinea coast, west Africa

    c. southeast China

    d. northern Australia

1. **(a)** Boothia Peninsula, which is almost an island, in Canada's Northwest Territories, is the northernmost point of the North American mainland. The North Magnetic Pole was once on its western shore.

2. **(c)** The North Atlantic islands of Greenland (the world's largest island) and Iceland are separated by Denmark Strait.

3. **(b)** An isthmus, such as the Isthmus of Panama, is the narrow landform with water on both sides that connects 2 larger land areas.

4. **(c)** The Strait of Hormuz connects the Persian Gulf with the Gulf of Oman.

5. **(b)** The Red Sea, which is about 1,200 miles long, is an inland sea between northeast Africa and the Arabian Peninsula.

6. **(a)** Quebec, 594,861 square miles, has the largest land area of all the Canadian provinces.

7. **(d)** The Great Barrier Reef, about 1,250 miles long off the eastcoast of Australia, is the largest deposit of coral in the world.

8. **(c)** Hoba West, the largest known meteorite (66 tons), was found in what is now Namibia, in Africa, in 1920.

9. **(b)** The Lascaux cave complex in southwest-central France was discovered by accident in 1940. Its powerful Paleolithic paintings in 13 different styles date from 150 centuries ago.

10. **(c)** The Gulf of Bothnia, the northern arm of the Baltic Sea, is between Finland on the east and Sweden on the west.

11. **(b)** Manhattan, Singapore, Lagos, and Montreal are all on islands.

12. **(a)** The opening of the Panama Canal in 1914 cut 7,800 miles from the sea voyage between New York and San Francisco. Sailing a vessel around South America can cost 10 times more in fuel and salaries. (Both the Panama and Suez Canals are too narrow for some modern vessels, such as supertankers.)

13. **(c)** The Strait of Gibraltar, between Africa and Spain, is 8 miles wide at its narrowest part.

14. **(a)** Rias are drowned rivers. Chesapeake Bay, on the Atlantic coast of the United States, is one of the world's largest rias. It is the drowned mouth of the Susquehanna River.

15. **(c)** The Rock of Gibraltar is 1,398 feet high. The British colony of Gibraltar is 2.5 square miles. Spain's claims to the colony have been rejected repeatedly by Great Britain.

16. **(c)** Vasco da Gama passed South Africa's Cape of Good Hope on his voyage to India in 1497, the first European skipper to make the trip. The cape, which is on the southwest coast of Cape Province, in the Republic of South Africa, is about 30 miles south of Cape Town. When Bartholomeu Dias rounded the cape in 1488, he dubbed it the Cape of Storms.

17. **(c)** Pomerania is a historical region on the Baltic Sea in northern Europe.

18. **(c)** The Arabic term *wadi* means "valley"; it can also mean "river" and "dry river bed" in place names in northern Africa and the Near East.

19. **(a)** Cape Disappointment is on the north side of the entrance to the Columbia River in southwest Washington.

20. **(d)** The world's largest contiguous plantations are managed by Firestone Rubber Company, at Harbel, the second largest city in Liberia, in western Africa.

21. **(a)** Angel Falls, in southeast Venezuela, is 3,212 feet high. It is on the side of a 20-mile-long flat-topped mountain east of the Caroni River.

22. **(d)** The Bering Sea and the Sea of Okhotsk are separated by the Kamchatka Peninsula, 750 miles long, in northeastern Soviet Union.

23. **(c)** The longest undefended border in the world is the frontier between the United States and Canada: 5,527 miles. The boundary was settled upon in 1783, after America's War of Independence from England, and by treaties in 1818, 1842, and 1846.

24. **(a)** Western Africa along the Bight of Benin on the Atlantic Ocean was long the Slave Coast.

25. **(d)** Japan's Seikan rail tunnel is about 33 miles long.

26. **(c)** The largest known single copper-mining property in the world is Chuquicamata, in northern Chile.

27. **(a)** Eritrea is Ethiopia's access to the Red Sea, in northeast Africa.

28. **(c)** The Silk Route, the ancient, overland trade roads between China and the Mediterranean world, was also called the Jade Road, the Emperor's Road, and the Fur Road.

29. **(b)** Vladivostok, which was founded in 1860, is the eastern terminus of the Trans-Siberian Rail Road and the principal Soviet seaport on the Pacific Ocean. The Soviet Union's Pacific fleet is headquartered there. In the winter, the harbor is kept open by icebreakers.

30. **(a)** The Bosporus, or Istanbul Strait, is the narrow waterway near Istanbul that separates Turkey in Asia from Turkey in Europe.

31. **(b)** Recife, a leading port of Brazil, with a population of more than a million, is the easternmost point of South America.

32. **(b)** Ancient geographers reckoned longitude by the Canary Islands' Hierro, which they thought was the western limit of the world.

33. **(d)** The sparsely populated Barren Grounds of northern Canada are low-level treeless plains with thin soil, swamps, and lakes.

34. **(d)** Rotterdam, or Europoort, handles more cargo than any other seaport in the world.

35 **(c)** First sighted by Sir Francis Drake, in 1578, Cape Horn on Horn Island projects south into Drake Passage in the southern Tierra del Fuego archipelago at the southern extremity of South America. Dutch navigators in 1616 named the cape after Hoorn, a town back home in Holland.

36. **(c)** The 550-mile-long Arabian Gulf is also called the Persian Gulf.

37. **(d)** Nicaragua's 300-mile-long swampy "Mosquito Coast" is on the Caribbean Sea.

38. **(b)** The northern African coastal region from the Atlantic Ocean to Egypt became the independent Muslim states known as Barbary States: Morocco, Algiers, Tunis, and Tripoli.

39. **(c)** Transylvania is a triangular 1,000-to-1,600-foot-high, 21,297-square-mile plateau in northwest and central Romania, in southeastern Europe.

40. **(a)** In the Second World War, the Hitler Line was a German defense line, a support for the Gustav Line, in western Italy. During the First World War, the German line of defensive fortifications established across northeast France in 1916 was the Hindenburg Line.

41. **(b)** White Man's Grave was the name formerly applied to the lands along the Guinea coast, in west Africa, more especially to the district around Freetown, capital of Sierra Leone. It is hot and humid there, the drier and healthier harmattan winds from the northeast rarely reaching the area.

1. Kosciusko is the highest mountain in
    a. Australia
    b. Central America
    c. Poland
    d. the Arctic

2. Eight of the world's 10 highest mountain peaks are in the
    a. Australian Alps
    b. Andes
    c. Himalayas
    d. Urals

3. The world's longest mountain chain is the
    a. Himalayas
    b. Andes
    c. Alps
    d. Rockies

4. Victoria Peak is the highest point on
    a. the British Isles
    b. the Nile River
    c. the island of Hong Kong
    d. the border of Afghanistan and Iran

5. The highest mountain in the Alps is in
    a. France
    b. Switzerland
    c. Germany
    d. Italy

6. Everest is the highest mountain in the world. The second highest mountain is
   a. K2
   b. Communism Peak
   c. Kilimanjaro
   d. Aconcagua

7. Africa's highest mountain soars _____ feet.
   a. 8,741
   b. 17,710
   c. 19,340
   d. 22,003

8. From base to peak, the tallest mountain in the world would be
   a. Mauna Kea
   b. Elbrus
   c. K2
   d. Gunnbjorn

9. South America's—and the western hemisphere's—highest mountain is _____ feet lower than the world's highest mountain, Everest, which stands 29,028 feet in the Himalayas.
   a. 1,073   c. 6,194
   b. 1,496   d. 10,850

10. The height of a mountain is considered to be the measurement between
    a. mean sea level and the summit
    b. its snowline and its peak
    c. its circumference and its peak
    d. its tree line and its peak

11. The highest point in Europe is
    a. Mont Blanc
    b. Mount Rysy
    c. Mont Triglav
    d. Mount Elbrus

12. Mountainous Norway's loftiest mountain is Glittertind, at
    _____ feet.
    a. 8,110
    b. 13,855
    c. 14,550
    d. 19,333

13. The Alps cross the borders of _____ European countries.
    a. 3
    b. 5
    c. 7
    d. 12

14. The _____ mountain range separates the 3 Soviet republics of
    Georgia, Armenia, and Azerbaijan from the rest of the Soviet Union.
    a. Ural
    b. Himalaya
    c. Caucasus
    d. Kjolen

15. The United States Range is a mountain range in
    a. Canada
    b. Africa
    c. Australia
    d. Antarctica

16. The Soviet Union's highest mountain is
    a. Communism Peak
    b. Narodnaya
    c. Denezhkin Kamen
    d. Yamantau

17. A total of _____ peaks of the Rocky Mountains in the state of Colorado soar higher than 14,000 feet.
    a. 5
    b. 10
    c. 50
    d. 125

18. Australia's principal mountain chain is
    a. the Great Dividing Range
    b. the Murchison Range
    c. the Robson Range
    d. the Gregory Range

19. Mount Everest, the world's highest mountain, straddles the border between
    a. Nepal and the People's Republic of China
    b. Afghanistan and the Soviet Union
    c. India and Pakistan
    d. Chile and Argentina

20. Canada's highest mountain is
    a. Mount Logan
    b. Mount Medicine Hat
    c. Mount Waddington
    d. Mount Albert

21. The highest mountain in North America is named after this President of the United States:

    a. Herbert Hoover

    b. Theodore Roosevelt

    c. Benjamin Harrison

    d. William McKinley

1. **(a)** Kosciusko, 7,316 feet, is the highest mountain in Australia. It is in the Australian Alps in the southeast corner of the continent.

2. **(c)** Eight of the world's 10 highest mountain peaks, including the world's highest, Mount Everest's, at 29,028 feet, are in the Himalayas.

3. **(b)** The Andes, the world's longest mountain chain, runs more than 4,000 miles along the west coast of South America.

4. **(c)** Victoria Peak, 1,805 feet, is the highest point on the 29-square-mile island of Hong Kong, part of the British crown colony of Hong Kong off southeast China.

5. **(a)** The highest mountain in the Alps is in France: Mont Blanc's peak is 15,781 feet.

6. **(a)** K2, also called Godwin Austen or Dapsang, the highest peak (28,250 feet) in the Karakoram Range, in a region controlled by Pakistan, is the second highest mountain in the world. It was first scaled in 1954.

7. **(c)** Mount Kilimanjaro is Africa's highest mountain, 19,340 feet.

8. **(a)** From its base, in the Pacific Ocean, to its peak over south-central Hawaii Island, the volcanic Mauna Kea would be the tallest mountain in the world: 33,476 feet; but only 13,680 feet are above mean sea level, and that's what counts.

9. **(c)** Mount Aconcagua, South America's—and the western hemisphere's—highest mountain, in western Argentina, stands at 22,834 feet, which is 6,194 feet lower than Everest.

10. **(a)** Because the height of a mountain is considered to be the measurement between mean sea level and the summit of the mountain, Pikes Peak, in Colorado, is therefore considered to be 14,110 feet high—its summit is that far above sea level but it rises only about 9,000 feet above the surrounding plains.

11. **(d)** The Soviet Union's Mount Elbrus, 18,510 feet high in the northern subsidiary spur of the main range of the Caucasus, is the highest point in Europe.

12. **(a)** At 8,110 feet, in the Jotunheimen, in the south-central part of the country, Glittertind is Norway's highest peak.

13. **(c)** The Alps cross the borders of 7 countries: Austria, France, Germany, Italy, Liechtenstein, Switzerland, and Yugoslavia.

14. **(c)** The Soviet republics of Georgia, Armenia, and Azerbaijan are separated by the Caucasus mountain range from the rest of the Soviet Union.

15. **(a)** The United States Range is a mountain range in the Northwest Territories, in northern Canada; the highest point is 9,000 feet.

16. **(a)** Communism Peak, 24,854 feet, is the Soviet Union's highest mountain. It is in the Pamirs, which contain some of the most challenging alpine treks in the world.

17. **(c)** Fifty peaks of the Rocky Mountains that are in Colorado soar higher than 14,000 feet.

18. **(a)** The Great Dividing Range, Australia's principal mountain chain, hugs the continent's eastern and southeastern coasts for about 2,000 miles. In the south, it disappears under Bass Strait, then reappears 130 miles from the mainland state of Victoria to form the 26,383-square-mile island state of Tasmania.

19. **(a)** Mount Everest, 29,028 feet, the world's highest mountain, straddles the border between Nepal and the People's Republic of China.

20. **(a)** Canada's highest mountain, Mount Logan, 19,523 feet, is in southwest Yukon Territory.

21. **(d)** Mount McKinley, 20,320 feet, the highest mountain in North America, is in Denali National Park, in south-central Alaska, and was named for the twenty-fifth US President, William McKinley, who was assassinated in Buffalo, New York, in 1901.

- It was to obtain plants for a botanic garden in the West Indies that Captain William Bligh made his famous voyage on the *Bounty* to Tahiti in the late 1780s.
- Norway's Atlantic coastline is approximately 1,500 miles long, but the fjords and islands add another 10,000 miles to the total.
- As the Aral Sea shrinks, the climate in Soviet Central Asia becomes hotter. It was once the fourth largest sea in the world, the size of Ireland. Its waters have been diverted to irrigate rice and cotton fields.
- With the rise of Tokyo to power in the early seventeenth century, the Japanese city of Kamakura saw its population fall from about a million to a few thousand.
- North Pole and South Pole, which are actually mere points without any dimension, have the same dimension as the Equator on the most used map of the world, the Mercator, designed in the sixteenth century by a Flemish geographer. The Mercator has great distortions in area and distance.
- The weight of the planet's atmosphere is equal to the weight of a layer of water 34 feet deep covering the Earth.
- The Hawaiian Islands consist mainly of the tops of a submerged volcanic mountain chain.
- Carbon dioxide, essential to plant life, makes up less than .04 percent of the planet's air.
- Europe's Alps and Africa's Atlas Range were formed at the same time. The Atlas Mountains rim the continent's northern edge for 1,500 miles.
- The southwest Pacific island of Nauru, whose economy has been based on phosphate sales, has no taxes.
- Magellan's name for the Philippines was Islands of Saint Lazarus.
- Turtle Island was what many Native Americans called what is today known as the North American continent.
- In the Texas panhandle, in the southwest United States, about 82,000 acres were planted to wheat in 1909. Two decades later, the acreage was 2 million. The wheat fields were "gold mines." But then it stopped raining, and "wheat heaven" became the nightmarish Dust Bowl.

- The Soviet Union claims just about 100 percent literacy.
- The republic of Sierra Leone, on the Atlantic in west Africa, has a 24-percent literacy rate, a male life expectancy of 35, and a female life expectancy of only a year longer.
- The "redwoods" of Antarctica are 100-year-old mosses. An American sealer captured the continent's sense of desolation: "When Mother Nature fashioned these . . . methinks she must have been drinking." Yet 65 million penguins and 35 million seals live there.
- The largest nation wholly in eastern Europe, Poland, 121,000 square miles, is the sixth largest in all of Europe.
- Per-acre yields in southern Europe are about half of western Europe's.
- There is a higher proportion of skilled labor in Lebanon than in any other Arab country.
- Water is essential for life, but water is not distributed evenly on Earth.
- Rwanda's population density is the highest in sub-Saharan Africa: 698.2 per square mile.
- Soviet astronomers have used the village of Pulkovo, 11 miles south of Leningrad, as the base for measurements instead of Greenwich, England, Prime Meridian for everyone else.
- Soils deposited by flowing water, such as in the Mississippi River valley, generate important food-producing areas.
- Only 3 percent of the US population lives in rural areas.
- The Florida and Korean peninsulas are about the same size.
- Because the air there is about 35 percent thinner than it is at sea level, people who grow up in La Paz, Bolivia, in the Andes, have larger-than-normal lungs. Food in La Paz cooks slower, and cocktails have more kick.
- North America has the longest coastline of any continent.
- Far northern North America stretches halfway around the world. Far southern Panama's narrowest section is only 31 miles across.
- The average Bangladeshi is only 16 years old.
- Costa Rica abolished its army in 1949.

- Luis Vaz de Camões's epic poem in 10 cantos, *The Lusiads*, is about the discovery by Vasco da Gama in 1497–98 of the sea route from western Europe to India, via southern Africa.
- There are about a half-million tremors and quakes in the Earth every year.
- Indonesia is the largest nonfederated nation state in the world.
- Eighty-seven percent of Brazil's agricultural land area is in holdings of more than 50 acres, a typical pattern of land ownership in Latin American countries.
- It was the long-standing friction between Hindus and Muslims that caused the creation of 2 countries out of British India: India for Hindus and Pakistan for Muslims.
- Cumana, a seaport city in northern Venezuela that has suffered often from earthquakes, is the oldest existing European settlement in South America.
- The oldest settlement of the Aleutian Islands, Unalaska, was established by Russians in the period 1760–65.
- The northern coastline of the Soviet Union extends about 3,000 miles on the Arctic Ocean.
- The former name of Uruguay was Banda Oriental.
- Bangladesh's maximum elevation is only 660 feet.
- In 1990, the United States had more than 2,000 Navy and Marine personnel on duty at Guantanamo, Cuba.
- The maximum depth of the largest lake in central Europe is 35 feet: Lake Balaton, 232 square miles, in western Hungary.
- Popularly considered the most northerly point of Great Britain is John o'Groat's House, a site in northern Scotland.
- Alexander Selkirk, the original of Daniel Defoe's heroic Robinson Crusoe, lived for 5 years in the early 1700s on Mas a Tierra, an island in the Juan Fernandez chain, in the South Pacific Ocean, about 400 miles west of Chile.

- Before the Second World War, Java produced almost all of the world's supply of quinine.
- The known plant species total at least 300,000.
- The eucalyptus tree, native to Australia, is widely used for reforestation elsewhere.
- The Caribbean island Antigua has so many droughts because it doesn't have enough trees to attract rainfall.
- In North Carolina is the highest point in the United States east of the Mississippi River: Mount Mitchell, 6,684 feet.
- One out of every 5 Australians is foreign-born.
- The most northerly land in the eastern hemisphere is the Franz Josef archipelago, about 8,000 square miles and about 190 islands, in the Arctic Ocean.
- There is only one place in the United States where the boundaries of 4 states come together: the point of intersection of Colorado, Utah, Arizona, and New Mexico is "Four Corners," at 37° N and 109° W.
- A temporary state called Franklin was organized in western lands of North Carolina (now part of eastern Tennessee) in 1784. It ceased to exist in February 1788.
- The "smallest city in the United States" is Vergennes, Vermont, with a population just over 2,000.
- The mining and fur-trading town of Verkhoyansk in the Soviet Union has a January mean temperature of minus 59° F. Political exiles have been sent there.
- About 1 million Dominicans, from the Dominican Republic, live in the United States, most of them in New York City. The population of the Dominican Republic is 6.5 million.
- India's armed forces are the fourth largest in the world.
- The Soviet Union's expanding agricultural lands to the south and east are subject to chronic drought, because 80 percent of the nation's river water courses northward into the remote Arctic Ocean.

1. The Lacandona rain forest is the largest tropical rain forest in
   a. North America
   b. Africa
   c. Asia
   d. Australia

2. All of the world's natural sodium nitrate, which is used to make explosives and fertilizer, comes from
   a. Siberia
   b. the Central African Republic
   c. Chile
   d. Bulgaria

3. About 30 percent of the noncommunist world's minable uranium reserves is in
   a. Japan
   b. Africa
   c. Antarctica
   d. North America

4. _____ is the world's leading producer of coal.
   a. The People's Republic of China
   b. The Soviet Union
   c. Australia
   d. The United States

5. The country with the largest forested area is
   a. Borneo
   b. Irian Barat
   c. Thailand
   d. the Soviet Union

6. The world's largest producer and exporter of silver is
    a. Chile
    b. Mexico
    c. the United States
    d. the Caroline Islands

7. Platinum is more valuable than gold. The number-one producer of platinum is
    a. Canada
    b. South Africa
    c. the Soviet Union
    d. Indonesia

8. The oldest known living tree is in
    a. the United States
    b. Romania
    c. Indonesia
    d. Finland

9. There are more known natural gas reserves in _____ than in any other country of the world.
    a. the Soviet Union
    b. Iran
    c. Algeria
    d. the United States

10. The fossil fuel coal formed from _____ that died millions of years ago.
    a. plants
    b. dinosaurs
    c. whales
    d. insects

11. The 3 principal natural resources of the Bahamas—a series of long, flat coral islands in the western Atlantic Ocean extending from about 50 miles off the southeast coast of Florida in an arc to the northern edge of the Caribbean—are salt, timber, and aragonite. Aragonite is
   a. red sand
   b. a safe form of asbestos
   c. sugar cane
   d. a type of limestone with several industrial uses

12. The world's supply of the pungent spice called clove has come exclusively from
   a. Indonesia
   b. Luzon, in the Philippines
   c. Sri Lanka
   d. Zanzibar

13. The location of many large cities in Great Britain was influenced by the availability of this natural resource:
   a. oil   c. plutonium
   b. coal   d. nickel

14. From each ton of gold ore comes
   a. 1/16 of an ounce of gold
   b. exactly the amount of 12 ounces of gold
   c. more silver than gold
   d. about half a ton of gold

15. Most of the planet's fresh water is stored in
   a. rocks
   b. craters of volcanoes
   c. glacial ice
   d. lakes and inland seas

16. _____ is the world's leading producer of crude oil.
  a. The Soviet Union
  b. Saudi Arabia
  c. Indonesia
  d. Colombia

17. Experts agree that _____ may possess as much as 25 percent of all of the oil remaining in the world.
  a. the Soviet Union
  b. Colombia
  c. Saudi Arabia
  d. the Bering Sea

18. The most plentiful metallic element in the crust of the planet is
  a. silver
  b. aluminum
  c. nickel
  d. molybdenum

19. Peat, a fossil fuel considered to be the first stage in the long transformation of plant material into coal, is generally found in _____ climates.
  a. cool, northern
  b. blistering hot, equatorial
  c. ice-cold
  d. benign

20. About a billion dollars' worth of emeralds are extracted from the Earth every year, half of them from
  a. Colombia
  b. Zambia
  c. Brazil
  d. the Soviet Union

21. The availability nearby of the natural resources coal and iron ore sparked the early growth of these 2 industrial American cities:
    a. Pittsburgh, Pennsylvania, and Birmingham, Alabama
    b. Chicago, Illinois, and Detroit, Michigan
    c. Cleveland, Ohio, and Cincinnati, Ohio
    d. Knoxville, Tennessee, and Lexington, Kentucky

22. Styrofoam and many other types of plastic are made by chemicals derived from this fossil fuel:
    a. petroleum   c. coal
    b. peat   d. natural gas

23. The world's largest rain forest is in
    a. South America
    b. Africa
    c. northwestern Canada
    d. Indonesia

24. The most important anthracite coal region of the United States is the state of
    a. Pennsylvania
    b. Ohio
    c. Virginia
    d. Minnesota

25. Petroleum and natural gas are the only significant natural resources of Bahrain, a group of islands in the Persian Gulf midway between the tip of the Qatar Peninsula and mainland Saudi Arabia. At the present rate of production, Bahrain's oil reserves will last another _____ years.
    a. 2   c. 75
    b. 10 to 15   d. 1,000

26. Long ago, this Mediterranean island was famous for its copper:
    a. Crete
    b. Sardinia
    c. Malta
    d. Cyprus

27. Large tracts of the Amazon tropical rain forest have been cleared for cattle ranches, for farms, for lumber, and
    a. for tax incentives
    b. for construction of a second capital city
    c. for national parks and entertainment centers
    d. for getting rid of deadly viruses

1. **(a)** North America's largest tropical rain forest, the Lacandona, in the southeast Mexican state of Chiapas, covered 5,000 square miles (the size of Connecticut) 50 years ago. More than 60 percent of the lush but fragile forest has been lost since 1970. The forest is named for a nation of Native Americans believed to have descended from the Mayas and to have inhabited the region since pre-Columbian times.

2. **(c)** The Atacama Desert, which runs 600 miles along the Pacific coast of Chile, is the source of all of the world's natural sodium nitrate.

3. **(d)** North America has about 30 percent of the noncommunist world's minable uranium reserves.

4. **(a)** The world's leading producer of coal is the People's Republic of China, which also has the largest coal reserves. (In the United States, Kentucky is the foremost state in coal production.)

5. **(d)** The Soviet Union has the world's largest forested area.

6. **(b)** Mexico is the world's largest producer and exporter of silver.

7. **(c)** More platinum is produced by the Soviet Union than by any other country. South Africa and Canada rank second and third, respectively, in production of platinum.

8. **(a)** The world's oldest known living tree (4,700 years) is a bristlecone pine called Methuselah on the California side of the White Mountains.

9. **(a)** The Soviet Union has the most known natural gas reserves in the world. Iran has the second most, the United States the third, Algeria the fourth.

10. **(a)** Plants that died millions of years ago became the dark fossil fuel coal, the most abundant of all fossil fuels. Coal is found on every continent.

11. **(d)** Aragonite is a type of limestone with several industrial uses.

12. **(d)** Cloves were the agricultural commodity with which the nineteenth-century prosperity of the southeast African island of Zanzibar was associated. Zanzibar united with Tanganyika in 1964 to form Tanzania.

13. **(b)** The availability of coal influenced the location of many large cities in Great Britain.

14. **(a)** From each ton of gold ore comes 1/16 of an ounce of gold.

15. **(c)** Antarctica locks up some two thirds of all fresh water on the planet. Other glaciers and ground water account for most of the rest of the fresh water.

16. **(a)** The world's leading producer of crude oil is the Soviet Union.

17. **(c)** Saudi Arabia, according to experts, may possess as much as one quarter of all of the oil remaining in the world.

18. **(b)** Aluminum is the most plentiful metallic element in the Earth's crust.

19. **(a)** The fossil fuel peat is generally found in cool, northern climates.

20. **(a)** About half of the world's emeralds that come out of the Earth every year are from Colombia. Zambia mines 20 percent, Brazil mines 15 percent, and even smaller amounts come from Zimbabwe, Pakistan, Afghanistan, Australia, Madagascar, Tanzania, and the Soviet Union.

21. **(a)** The early growth of both Pittsburgh, Pennsylvania, and Birmingham, Alabama, can be attributed to the availability of the nearby natural resources coal and iron ore.

22. **(a)** Chemicals derived from petroleum are used to make plastics.

23. **(a)** The Amazon in South America is the world's largest rain forest. It is an area almost as big as 48 of the 50 states of the United States.

24. **(a)** Pennsylvania is the most important anthracite coal region in the United States.

25. **(b)** Bahrain, the first Arab state with an oil refinery, has oil for another 10 to 15 years at the present rate of production. Its gas reserves should last for another half century.

26. **(d)** The east Mediterranean island of Cyprus was famous long ago for its copper.

27. **(a)** Tax incentives also prompt clearing of the Amazon rain forest.

DON'T BLOW IT —

GOOD PLANETS ARE HARD TO FIND

— Bumper Sticker

1. There are more Greek-speaking people in _____ than in any other city outside Greece.
   a. Rome, Italy,
   b. Cairo, Egypt,
   c. Boston, Massachusetts,
   d. Melbourne, Australia,

2. More than 20 percent of the population of Australia lives in the city of
   a. Melbourne
   b. Alice Springs
   c. Sydney
   d. Perth

3. The Tamils and the Sinhalese are the 2 major ethnic groups of
   a. Afghanistan
   b. Sri Lanka
   c. Ghana
   d. Turkey

4. Apartheid became an official policy of the government of South Africa in
   a. 1898
   b. 1919
   c. 1934
   d. 1948

5. The official language of all but 3 of the nations of South America is
   a. English
   b. Spanish
   c. Portuguese
   d. French

6. Peru's largest ethnic group is its _____ population.
   a. Portuguese
   b. Native American
   c. Filipino
   d. Italian

7. The South African city of Johannesburg (population about 1,609,000) owes its development to
   a. uranium
   b. the slave trade
   c. gold
   d. apartheid

8. There are more Hindus in _____ than in any other country of the world.
   a. Vietnam
   b. India
   c. Taiwan
   d. Egypt

9. There are more people of Asian descent in _____ than in any of the other 50 states of the United States.
   a. Hawaii
   b. California
   c. Texas
   d. New York

10. Swahili is spoken throughout
    a. east Africa
    b. Iran
    c. Saudi Arabia
    d. Pakistan

11. Thirty-seven percent of the population of this South American
     nation is East Indian, 31 percent is Creole, 15 percent is Javanese,
     10 percent is black, and 2 percent is Chinese:
     a. Uruguay
     b. Paraguay
     c. Suriname
     d. Venezuela

12. _____ claims the highest literacy rate in the Arab world.
     a. Bahrain
     b. Yemen
     c. Oman
     d. The United Arab Emirates

13. The fastest growing ethnic element in the population of
     Czechoslovakia is
     a. Gypsies
     b. Serbs
     c. Hungarians
     d. Arabs

14. About _____ percent of the population of the 700-island
     Bahamas, in the western Atlantic Ocean off the southeast coast of
     Florida, is black.
     a. 1   c. 50
     b. 20   d. 85

15. Since 1980, more refugees have fled from _____ than from any
     other country in the world.
     a. the Philippines
     b. the Soviet Union
     c. Afghanistan
     d. Romania

16. During the 3 centuries of the Africa-to-the-New World slave trade, some _____ Angolans from the west coast of southern Africa are estimated to have been shipped across the Atlantic Ocean.

   a. half a million

   b. 2 million

   c. 3 million

   d. 5 million

17. In the landlocked Himalayan country of Bhutan, which has a population of about 1.3 million, suffrage is

   a. 1 vote per family

   b. for every family member, no matter the person's age

   c. only for men who have lost at least 2 kinsmen in war

   d. only for women over 50 years of age

18. Since 1960, the greatest number of Asian immigrants in the United States have come from

   a. Vietnam

   b. Cambodia

   c. Hong Kong

   d. the Philippines

19. India's most populous city is

   a. Bombay   c. Madras

   b. Calcutta   d. Kanpur

20. During the Second World War, about 1.3 million American soldiers crossed the Atlantic Ocean and debarked in

   a. Oran, North Africa

   b. Gourock, Scotland

   c. Gotham, in Nottinghamshire, England

   d. Stockholm, Sweden

21. India's urban population today is about 180 million. By the year 2025, its urban population is expected to be about
   a. 150 million
   b. 250 million
   c. 350 million
   d. 660 million

22. The 10 most populous countries in the world account for _____ of the world's population.
   a. one tenth
   b. one quarter
   c. one half
   d. two thirds

23. The northernmost city in Europe has uninterrupted daylight May 17 to July 29 but residents there don't see the Sun from November 21 to January 21. The city is
   a. Glasgow
   b. Hammerfest
   c. Archangel
   d. Murmansk

24. There are fewer infant deaths per 1,000 live births in the first year of life in _____ than in any other country of the world.
   a. Finland
   b. Iceland
   c. Japan
   d. Sweden

25. Nine African nations suffered massive drought in the first half of the 1980s. The most seriously affected nation was
   a. Chad
   c. Egypt
   b. Sudan
   d. Ethiopia

26. Outside of Japan, there are more Japanese in _____ than in any other city.
    a. São Paulo, Brazil,
    b. Los Angeles
    c. Sydney, Australia,
    d. Seoul, Korea,

27. The United States, 3,615,123 square miles, is not as large as its northern neighbor, Canada, 3,851,809 square miles, but it has _____ times as many people.
    a. 3
    b. 4
    c. 7
    d. 10

28. About _____ languages are spoken throughout Africa.
    a. 17
    b. 27
    c. 73
    d. 800

29. Persons of mixed Spanish or Portuguese blood and Indian blood are
    a. mestizos
    b. Hispanics
    c. Fijians
    d. East Indians

30. The human population of the world is approaching
    a. 500 million
    b. 4.5 billion
    c. 5.5 billion
    d. 12 billion

31. Western Europe's most populous country is
    a. Germany
    b. the Netherlands
    c. Spain
    d. Portugal

32. The tropics encompass 36 percent of the Earth's land. About
    _____ of the world's people live in the tropics, which are
    between the parallels of latitude called the Tropic of Cancer and the
    Tropic of Capricorn.
    a. one quarter
    b. one third
    c. one half
    d. three quarters

33. About _____ million people live in the 40 countries of black
    Africa.
    a. 5
    b. 15
    c. 180
    d. 300

34. The most populous Muslim countries are in
    a. Asia
    b. eastern Europe
    c. eastern Africa
    d. South America

35. The homeland of the Basques is in the
    a. Pyrenees
    b. Andes
    c. Southern Alps
    d. Himalayas

36. The Roman Empire in the Old World and the Incan Empire in the New World were both notable for
    a. human sacrifices to weed out the old and the handicapped
    b. polygamy
    c. roadbuilding
    d. gladiator tournaments

37. Long before Columbus "sailed the ocean blue," these Europeans had established settlements on the east coast of North America:
    a. Frenchmen
    b. Irishmen
    c. Germans
    d. Norsemen

38. _____ percent of Egypt's 53 million people live within a dozen miles of the Nile River or one of its delta distributaries.
    a. Seven
    b. Twenty-eight
    c. Sixty-three
    d. Ninety-five

39. About _____ percent of Canadians live in the southern part of their vast nation.
    a. 90   c. 95
    b. 93   d. 9

40. Canada's 2 territories—Yukon and the Northwest Territories—constitute more than a third of the area of the second largest country in the world, and they have about _____ percent of the country's population.
    a. 33   c. 15
    b. 25   d. 1

41. There are more foreign refugees in _____ than in any other country of the world.

   a. Israel

   b. the United States

   c. Cambodia

   d. Pakistan

42. Africa's most populous country is

   a. Egypt

   b. Nigeria

   c. Chad

   d. Ethiopia

43. The giant Asian subcontinent country of India occupies 2.2 percent of the world's land area. It is home to nearly _____ percent of the world's population.

   a. 3   **c.** 17

   b. 10   **d.** 26

44. It has been calculated that at least 100,000 people pay for the right to sleep on a stretch of sidewalk in

   a. Hanoi, Vietnam

   b. Bombay, India

   c. Bogotá, Colombia

   d. Dar es Salaam, Tanzania

45. The disease that killed about a third of Europe's population in the fourteenth century was

   a. malaria

   b. cholera

   c. the bubonic plague

   d. measles

46. In the 100 years between 1835 and 1935, perhaps as many as
_____ Europeans left their homes for other countries.

    a. 1.5 million

    b. 10 million

    c. 15 million

    d. 75 million

47. The population of the Earth first reached 1 billion in the
_____ century.

    a. second

    b. sixth

    c. early nineteenth

    d. mid-twentieth

48. More than a million and a half people live in Yangon, the capital of
Myanmar. Two centuries ago,

    a. there was not even a town on the site

    b. 5 million people lived there

    c. the city was repeatedly destroyed by earthquakes and floods

    d. there was evidence that aliens from another planet had landed
there

49. Other than Antarctica, which has no permanent population and no
cities, the population of this continent is the least urbanized:

    a. Africa

    b. Asia

    c. South America

    d. Australia

1. **(d)** There are more Greek-speaking people in Melbourne, Australia, than in any other city outside Greece.

2. **(c)** The city of Sydney is home to more than 20 percent of Australia's population.

3. **(b)** The 2 major ethnic groups of the Indian Ocean-island of Sri Lanka are the Tamils and the Sinhalese.

4. **(d)** It was in 1948 that apartheid became an official policy of the government of South Africa.

5. **(b)** Spanish is the official language of all but 3 of the 12 nations of South America. Brazilians speak Portuguese, Surinamese speak Dutch, Guyanese speak English.

6. **(b)** The largest ethnic group in Peru is Native American.

7. **(c)** Johannesburg is the South African metropolis that gold made.

8. **(b)** There are more Hindus in India than in any other country of the world.

9. **(b)** In the United States, California is home to the largest number of people of Asian descent.

10. **(a)** Swahili, which is spoken throughout east Africa, was influenced and spread by Arab traders.

11. **(c)** Suriname, formerly Netherlands Guiana, or Dutch Guiana, is sandwiched between Guyana and French Guiana in the northeast corner of the continent. Thirty-seven percent of the population is East Indian.

12. **(a)** Bahrain, which is an archipelago consisting of 33 islands in the Persian Gulf, only 5 of them inhabited, claims the highest literacy rate in the Arab world: about 74 percent. By contrast, the literacy rate of Yemen, on the Red Sea in the southwestern corner of the Arabian Peninsula, is only 20 percent.

13. **(a)** Gypsies represent the fastest growing ethnic group in the population of Czechoslovakia. There are about 300,000 in this east European country.

14. **(d)** Blacks make up 85 percent of the population of the Bahamas, which were a staging area for the slave trade. Many blacks were also taken there by thousands of British loyalists who fled the American colonies during the Revolutionary War.

15. **(c)** The Soviet Union's invasion of Afghanistan generated the most massive recent exodus of people from any country in the world.

16. **(c)** Around 3 million Angolans were enslaved and exported in the seventeenth, eighteenth, and nineteenth centuries; many of them were shipped to plantations in Brazil.

17. **(a)** Suffrage is 1 vote per family in Bhutan, the landlocked Himalayan country located between the Tibetan plateau and the Assam-Bengal plains of northeastern India.

18. **(d)** The Philippines were home to the largest total number of Asians who have moved to the United States since 1960.

19. **(b)** Calcutta, with more than 9 million people, is India's most populous city.

20. **(b)** It was at Gourock, on the south shore of the Firth of Clyde, in southwest Scotland, that about 1.3 million American soldiers debarked during the Second World War.

21. **(d)** It is expected that India's urban population in the year 2025 will have expanded to 660 million. The huge south Asian subcontinent is already the world's second-most populous country, second only to the more than 1 billion people of the People's Republic of China.

22. **(d)** Most nations have small populations. More than half of the world's countries have fewer people than the state of Virginia (which has a population of about 5.5 million, making it the 14th state in population size in the United States). The world's 10 most populous countries are home to two thirds of the world's population.

23. **(b)** Hammerfest, in northern Norway, is the northernmost city in Europe.

24. **(a)** Finland has the fewest infant deaths per 1,000 live births (all races) in the first year of life. Iceland has the second fewest, Japan the third fewest.

25. **(d)** In Ethiopia, the hardest hit of the 9 African nations that suffered massive drought in the first half of the 1980s, millions of people died.

26. **(a)** The largest Japanese community outside of Japan is in São Paulo, Brazil.

27. **(d)** The population of the United States, around 250 million, is 10 times the population of Canada.

28. **(d)** About 800 languages are spoken throughout Africa.

29. **(a)** Mestizos are persons of mixed Spanish or Portuguese and Indian blood. Most mestizos live in South America.

30. **(c)** There are between 5.2 billion and 5.5 billion people in the world.

31. **(a)** Germany is the most populous country in western Europe, with a population of about 80 million people.

32. **(b)** About one third of the world's population lives in the tropics. The Tropic of Cancer is the line of latitude about 23.5° N of the Equator. The Tropic of Capricorn is the line of latitude about 23.5° S of the Equator.

33. **(d)** It is estimated that 300 million people live in the 40 countries of black Africa.

34. **(a)** The world's most populous Muslim countries are India, Indonesia, Bangladesh, and Pakistan, in southern Asia and southeast Asia.

35. **(a)** The Basques, noted for their seafaring and sheepherding skills, live in the western Pyrenees Mountains of Spain and France.

36. **(c)** Two ancient empires, the Roman in the Old World and the Incan in the New World, were both notable for extensive road building.

37. **(d)** Norwegians, Vikings, Norse, Northmen—Norsemen!—had established settlements on the east coast of North America hundreds of years before Columbus "discovered" the New World. The Vikings may have gone as far inland as what is now Minnesota.

38. **(d)** Ninety-five percent of Egypt's 53 million people live within a dozen miles of the Nile River or one of its delta distributaries. It is estimated that if the Aswan Dam upstream were breached, almost every Egyptian would be drowned within 3 days.

39. **(a)** About 90 percent of Canadians live in the southern part of their vast nation.

40. **(d)** Less than 1 percent of Canada's 25 million population live in the Yukon and the Northwest Territories, the 2 provinces that constitute more than a third of the land area of the whole country.

41. **(d)** Pakistan houses more foreign refugees than any other country; more than a million escaped the Soviet Union's invasion of Afghanistan in the 1980s.

42. **(b)** There are about 85 million people in Africa's most populous country, Nigeria, which occupies 356,669 square miles of the western part of the continent.

43. **(c)** India is home to nearly 17 percent of the world's population.

44. **(b)** At least 100,000 people in Bombay pay for the right to sleep on a stretch of sidewalk.

45. **(c)** The bubonic plague, or the black death, or the black plague, killed about one third of Europe's population in the fourteenth century.

46. **(d)** The Americas, Africa, and Australia were the principal "better-life" destinations for the perhaps as many as 75 million Europeans who emigrated between the years 1835 and 1935.

47. **(c)** The population of the planet first reached 1 billion in the early nineteenth century. It took little more than another century for there to be 2 billion people. Between 1930 and 1975, the figure doubled again, to 4 billion. It is expected that the population of the world will be 6 billion by the end of this century and 8 billion by the year 2025.

48. **(a)** Where Yangon, Myanmar's capital, is today, no town existed 2 centuries ago.

49. **(a)** Africa is the least urbanized of the populated continents.

The Great Lakes Superior, Michigan, and Huron lie in scale over the United Kingdom of Great Britain.

- Adult literacy in Benin, on the south side of the west African bulge, is 11 percent. Life expectancy is under 50 years. Annual per-capita income: $374.

- By 1911, practically the entire continent of Africa—save for Ethiopia and Liberia—was under the domination of European countries. German East Africa prevented the British from realizing the dream of an unbroken stretch of empire from Cairo, Egypt, to Cape Town, South Africa.

- The floor of Africa's tropical rain forest is dark because the sun cannot penetrate the vegetation.

- Mali, formerly the French Sudan, and Niger, formerly French West Africa, are both four fifths the size of Alaska.

- Zimbabwe, formerly Southern Rhodesia, is the size of California.

- On June 15, 1908, the United States reserved all unclaimed land lying within 60 feet of the US-Canadian boundary.

- The last land added to the lower 48 states of the United States was the Gadsden Purchase of 1853: parts of New Mexico and Arizona.

- The Valley of Ten Thousand Smokes, a volcanic region in southwest Alaska, was formed at the eruption of Katmai in 1912. Its floor had millions of steam jets; the temperature of some was as high as 1,200° F.

- Van, the largest lake in Turkey, at 1,419 square miles, has no apparent outlet.

- Venice, Italy, is built on 118 islands in the Lagoon of Venice. The city has 400 bridges.

- Snow avalanches are the natural hazard that causes the most damage in both Colorado and Switzerland.

- Malaysia declared in April 1989 that it would not accept any more Vietnamese refugees.

- The "farthest city in the world" is Ushuaia, in Tierra del Fuego National Territory, in southern Argentina. Population: 11,000.

- The *altiplano*, as it is called in Spanish, or "high plateau," is a wind-swept, almost perfectly flat plateau about 400 miles wide where the Andes split into 2 ranges and Argentina, Bolivia, and Chile meet. It is nearly 2.5 miles above sea level, an unexpected sight nestled among towering peaks.
- Indonesia is part of a mountain chain that is mostly hidden under the sea.
- Over 2 million mainland Chinese fled their homes and settled in Taiwan in the late 1940s.
- More than half of Germany's woodlands, including part of the Black Forest, has been assaulted by acid rain and all but decimated.
- Ireland gets about 7 feet of rain annually.
- Antarctica has sloughed off an iceberg roughly the size of Belgium.
- Tropical forests create much of their own rain through transpiration and evaporation.
- Africa's first desalination plant was opened in 1969 in Nouakchott, the capital of Mauritania, in west Africa, where the population has zoomed from 12,000 in 1964 to more than 350,000 today. Drought has lowered the country's water level so much that only the very deepest wells produce. Most of Mauritania's livestock has starved to death.
- Nearly a quarter of the land surface of China is limestock rock.
- Aristotle called wind the "dry sighs of the breathing Earth."
- Crocodile-like reptiles once dozed in swampy shallows in Arizona.
- Trinidad, the largest island of the Lesser Antilles chain in the Caribbean, is separated from Venezuela only by the 7-mile strait of the Gulf of Paria.
- Iran's central region is one of the most arid regions on the planet.
- More than a million Palestinian Arabs reside in Jordan.
- Until 1867, Alaska was known as Russian America.
- The Rocky Mountains are an extension of the South American Andes.
- The main source of the world's cobalt is Zaire.

- There are Southern Alps. They extend almost the entire length of New Zealand's South Island. Many peaks are 8,000 to 11,000 feet, and the area is noted for its scenery.
- Indonesia has the world's largest Muslim population, but it is not an Islamic state.
- The Sahara, about 3.5 million square miles, ranges from 100 feet below sea level to more than 11,000 feet above.
- Humidity averages 82 percent throughout the year in Brunei Darussalam, a Malay Muslim monarchy on the northwest coast of Borneo, 265 miles from the Equator. Oil and natural gas production there has generated one of the highest per-capita incomes in the world. Brunei Shell is said to employ almost half of the labor force.
- Switzerland is almost totally lacking in raw materials. It depends on the economy of the world for its prosperity, and it has one of the highest standards of living in the world. In 1986 a national referendum rejected membership in the United Nations.
- Virtually no rain falls from May to September on the Mediterranean island of Malta, about 58 miles south of Sicily and about 180 miles from North Africa.
- The Andes occupy nearly 27 percent of Peru's land area.
- Women outnumber men in the Soviet Union to a greater extent than in any other country in the world.
- Nearly 90 percent of Jordan is a wasteland.
- More than 2 million ethnic Germans were expelled from Czechoslovakia after the Second World War.
- More than 800 US corporations, with investments of about $6 billion, are located in Hong Kong, which is the world's busiest international port in terms of containers handled.
- The Tigris and Euphrates Rivers carry about 70 million cubic meters of silt annually to the delta in the Persian Gulf.
- Some of the largest documented meteorite craters are in northern Canada.

- The deepest point that humans have reached beneath the Earth's surface is the bottom of Western Deep Levels Mine, in a goldfield near Johannesburg, South Africa: 11,736 feet. It takes the miners a half-hour elevator ride to reach the site.
- About 10,000 Icelanders and about half of the North Atlantic island's livestock were killed by a volcanic eruption in 1783.
- All of Chile's domestic petroleum comes from the Strait of Magellan and the island of Tierra del Fuego, both at the southern tip of the country.
- Twenty-five percent of the population of Uruguay is of Italian origin.
- The only place north of Panama in the western hemisphere from where on a clear day both the Atlantic and the Pacific Oceans can be seen is the volcano Irazu, in central Costa Rica.
- The United States is the largest investor in the Philippines, Malaysia, and Singapore, and the United States and Japan are the largest investors in Indonesia and Thailand.
- The world's largest natural asphalt bog, measuring 100 acres, is Pitch Lake, on the southwest coast of Trinidad.
- The most perfect volcanic cone known is Mount Mayon's, at 8,284 feet, in the Philippines. Its last destructive eruption was in 1928.
- Tens of thousands of Egyptians live in Cairo's "city of the dead"—amid tombs, gravestones, and monuments.
- At the current rate of production, known copper reserves in Zambia will be exhausted by the turn of the century.
- Land in Yemen once used for productive agriculture has been turned over to the production of the cash crop qat, a mild amphetamine chewed by Yemenis. It has no significant export market.
- At any given time, an estimated 200,000 workers are absent from Lesotho, an enclave within the east-central part of the Republic of South Africa. Most of them spend up to 9 months a year in mining, farming, or industry in South Africa.
- The Pilgrims lived for 11 years in Leiden, in southwest Netherlands, before they sailed for the New World in 1620.

- The Bronx—41 square miles—is the only one of the 5 boroughs of New York City that is on the mainland of the United States.
- About 50 percent of Mexico is deficient in moisture every year.
- Canada's highest waterfall, with a total drop of 1,650 feet (with its highest single fall 1,200 feet), is Takkakaw, in southeast British Columbia.
- The average temperature in Bangladesh is 84° F.
- When Stalin became the Soviet dictator in the 1920s, less than 20 percent of the nation was urbanized. Today, about 70 percent of the population lives in or around cities.
- Antarctica is the highest of the continents, with an average elevation of 8,000 feet.
- At one time, Norway, Sweden, Finland, the Faroe Islands, Iceland, and Greenland were all under Danish rule.
- The favorite stream of Izaak Walton (1593-1683), the best-known fisherman ever, was Dove River, in central England.
- Europe's largest inland river port (above tidewater) is Duisburg, on the Rhine River at the confluence of the Ruhr, 12 miles north-northwest of Düsseldorf, Germany.
- The longest river in the British Isles is the Shannon, about 240 miles, in Ireland.
- About 2 million fur seals make their home on Pribilof Island, in the southeast Bering Sea, Alaska.
- The Great Lakes were ordinary river valleys until ice gouged them to a depth of hundreds of feet.
- In the early 1980s, the Caribbean 2-island nation Trinidad and Tobago attained the third highest per-capita ranking in the western hemisphere (the United States and Canada were first and second), thanks to its petroleum boom.
- About 4 million Mongols live outside Mongolia, most of them in the People's Republic of China. Less than 1 percent of their nation is fit for agricultural cultivation.

- Canada's most mountainous province is British Columbia, on the Pacific coast.
- Probably the oldest human remains ever found in the western hemisphere were a 12,000-year-old skeleton discovered in central Mexico in 1947.
- About one half of the world's supply of asbestos comes from Thetford Mines, about 50 miles south of Quebec, Canada.
- Tibet is the world's highest region; it averages 16,000 feet above sea level. It has been a nominally autonomous region within the People's Republic of China since 1965.
- Kiritimati, one of the Line Islands in the central Pacific south of Hawaii, is the largest atoll in the Pacific: 234 square miles. (It was discovered in 1777 by Captain James Cook.)
- At high water in summer, Tung-t'ing, usually a 1,430-square-mile lake in southeast-central China, expands up to 4,000 square miles.
- The Sudan is the largest country in Africa—nearly a million square miles—and the Nile River flows the entire length of the country.
- One acre of a nutrient solution can yield more than 50 times the amount of lettuce grown in the same amount of untreated soil.
- Sigsbee Deep is the deepest point in the Gulf of Mexico: 12,425 feet, in the southwest-central part of the gulf.
- The highest peak in Ireland is Carrantuohill: 3,414 feet.
- The world's heaviest rainfall, averaging 457 inches a year, is at Cherrapunji, a former British military station, in northeast India.
- Chile is approximately 2,650 miles long, but nowhere is it more than 221 miles wide.
- The motto of the state of Oregon is "She Flies with Her Own Wings."
- The largest tract of swamp in Europe is the Pripet Marshes in the western Soviet Union. The marshes are nearly uninhabited and mostly impassable, except in winter, when they are frozen.
- South America's landlocked republic of Paraguay has only one large lake: Ypoa, about 100 square miles in area.

1. The largest state in the United States in area east of the Mississippi River is
    a. Pennsylvania
    b. New York
    c. Georgia
    d. Illinois

2. The fiftieth state to join the Union, the Hawaiian Islands, stretches _____ miles across the Pacific Ocean.
    a. 415
    b. 590
    c. 1,000
    d. 1,500

3. When the British explorer Captain James Cook discovered the Hawaiian Islands in the Pacific Ocean, in 1778, he named them the
    a. Garden of Eden Islands
    b. Sandwich Islands
    c. House of Lords Islands
    d. Admiralty Islands

4. The first state in area in the United States is _____ , and the fiftieth state in population is _____ .
    a. Alaska . . . Wyoming
    b. Alaska . . . Alaska
    c. Alaska . . . the Hawaiian Islands
    d. Texas . . . New Mexico

5. If the Hawaiian Islands do not count their Pacific Ocean realms, the US state with the greatest water area is
    a. Mississippi   c. Oregon
    b. Missouri   d. Minnesota

6. Two states in the United States each border on 8 other states. The 2 are
    a. Wyoming and Idaho
    b. Illinois and Indiana
    c. Oklahoma and Kansas
    d. Tennessee and Missouri

7. The _____ area has the largest Arab population in the United States.
    a. Dallas, Texas,
    b. Jersey City, New Jersey,
    c. Springfield, Vermont,
    d. Detroit, Michigan,

8. The motto of this state of the United States is "The Life of the Land Is Perpetuated in Righteousness":
    a. Oklahoma
    b. Nevada
    c. Arizona
    d. Hawaii

9. The highest point in the United States east of the Rocky Mountains is in the state of
    a. North Carolina
    b. Vermont
    c. New Hampshire
    d. South Dakota

10. In the United States, the largest tract of hardwood and the largest virgin forest of red spruce are found in the _____ part of the country.
    a. eastern   c. western
    b. southern   d. northern

11. Two Presidents of the United States were born in the Massachusetts community of
    a. Springfield
    b. Northampton
    c. Quincy
    d. Boston

12. There are about _____ known caves in the United States.
    a. 350
    b. 700
    c. 5,000
    d. 17,000

13. The deepest lake in the United States is
    a. Crater Lake
    b. Lake Michigan
    c. Great Salt Lake
    d. Lake Champlain

14. The most populous state has more people than Canada or Australia or three quarters of the countries of Europe:
    a. Texas
    b. Florida
    c. California
    d. New York

15. Eleven states of the United States are in the " _____ belt."
    a. rust
    b. corn
    c. frost
    d. borscht

16. _____ is the state of the 50 United States that has the highest percentage of Hispanics in its population.
    a. Florida
    b. New York
    c. Texas
    d. New Mexico

17. The nicknames of this state in the midwest United States include "Sucker State" and "Prairie State":
    a. Illinois
    b. Missouri
    c. Indiana
    d. Wisconsin

18. Black Americans made up _____ percent of the total population of the United States in the 1980 census.
    a. 7   c. 22.4
    b. 11.8   d. 29.9

19. Alsatians migrated to America in the early eighteenth century and settled on the "German coast," which is
    a. north of New Orleans, Louisiana
    b. an enclave in Ohio
    c. the south shore at the eastern end of Cape Cod, Massachusetts
    d. northern Maine

20. This President of the United States was particularly active in the early days of the conservation movement:
    a. Thomas Jefferson
    b. James K. Polk
    c. Theodore Roosevelt
    d. Warren G. Harding

21. In the United States, the state with the highest infant-mortality rate per 1,000 live births in the first year of life is

    a. South Carolina

    b. Mississippi

    c. Georgia

    d. Alabama

22. _____ states of the United States border on the Gulf of Mexico.

    a. Three

    b. Four

    c. Five

    d. Seven

23. In the United States, the state of _____ can lay claim to being "the mother of three seas."

    a. Alabama

    b. Wisconsin

    c. Minnesota

    d. North Dakota

24. The northernmost town in the United States is

    a. Nome, Alaska

    b. Barrow, Alaska

    c. Orono, Maine

    d. Juneau, Alaska

25. British political control over their 13 colonies along the Atlantic seaboard was assured by

    a. the English defeat of the French in the French and Indian War (1754–1763)

    b. Russia's decision to remove its troops from North America

    c. the placing of nearly a million Redcoats in the colonies

    d. the Americans' strong desire to be ruled by England

26. The fiftieth state to join the Union, Hawaii, consists of _____ islands.

    a. 4

    b. 9

    c. 22

    d. 132

27. This midwestern city in the United States became "queen of the west" with the opening of the Miami and Erie Canal in 1832:

    a. Cleveland, Ohio

    b. Cincinnati, Ohio

    c. Louisville, Kentucky

    d. Gary, Indiana

28. New York-New Jersey has been replaced by _____ as the busiest port in the United States.

    a. Baltimore

    b. Charleston

    c. New Orleans

    d. Los Angeles

29. From this major city in the 48 contiguous United States one can drive south to enter Canada:

    a. Butte, Montana

    b. Buffalo, New York

    c. Niagara, New York

    d. Detroit, Michigan

30. The lowest point and the highest point in the contiguous 48 states of the United States are _____ miles apart.

    a. 80   **c.** 1,888

    b. 1,328   **d.** 2,004

31. A city in the United States with a population of half a million is
    located exactly 5,280 feet above sea level and is called "the mile-
    high city." The city is
    a. Spokane, Washington
    b. Minot, North Dakota
    c. Denver, Colorado
    d. Santa Fe, New Mexico

32. The state on the Atlantic coast of the United States that is closest to
    California is
    a. Massachusetts
    b. North Carolina
    c. Delaware
    d. Georgia

33. It depends on whose statistics are used, but it appears that the
    United States ranks the _____ lowest in infant mortality
    among all nations of the world.
    a. third   c. seventeenth
    b. tenth   d. nineteenth

34. The Snake River in the northwestern United States is _____
    miles long.
    a. 75   c. 750
    b. 175   d. 1,038

35. The northernmost point in the United States is Point Barrow,
    Alaska, which is about 5,000 miles north of the Equator. The
    southernmost US territory is 1,000 miles south of the Equator:
    a. Rose
    b. Guam
    c. Wake Island
    d. the Hawaiian Islands

36. The state of the United States with the wettest climate is
    a. Hawaii
    b. Washington
    c. Minnesota
    d. Mississippi

37. Much of the southwest United States is
    a. desertlike
    b. a rain forest
    c. tundralike
    d. animal-free

38. The distance between the 2 points farthest apart in the 48
    contiguous states of the United States is _____ miles.
    a. 1,893
    b. 2,430
    c. 2,897
    d. 3,103

39. The Toledo War of the mid-1830s in the United States was a dispute
    between the states of _____ and _____ over the location of
    their common boundary.
    a. Ohio . . . Michigan
    b. Ohio . . . Indiana
    c. Ohio . . . New York
    d. Ohio . . . Kentucky

40. More than at any time since the First World War, the US population
    increase is being driven by
    a. education
    b. migration
    c. the weather
    d. terrorism

41. The 2 most important crops grown in Iowa, Illinois, and Indiana, in the midwest United States, are corn and
     a. potatoes
     b. soybeans
     c. millet
     d. barley

42. Tombstone, in the southeast corner of the state of _____ , was known principally for its lawlessness.
     a. Georgia
     b. Florida
     c. South Dakota
     d. Arizona

43. On the very same day as the historic great fire in Chicago—October 8, 1871—another city in the midwest United States was also practically destroyed by a fire:
     a. Peshtigo, Wisconsin
     b. Springfield, Missouri
     c. Gary, Indiana
     d. St. Paul, Minnesota

44. The Appalachian National Scenic Trail in the eastern United States originates in the state of Maine and ends in
     a. New York   c. North Carolina
     b. West Virginia   d. Georgia

45. Washington, D.C., the third capital of the United States, was created from portions of two states:
     a. Maryland and Delaware
     b. Maryland and Virginia
     c. Virginia and West Virginia
     d. New York and Pennsylvania

46. The motto of this north-central state of the United States is "The Crossroads of America":

   a. Indiana   **c.** Missouri

   b. Wisconsin   **d.** Illinois

47. Symbolic of the geographical division between the northern states and the southern states of the United States is

   a. the Ohio River

   b. 35° North latitude

   c. the nation's capital, Washington, D.C.

   d. the Mason-Dixon line

48. Hampton, on Hampton Roads in southeast Virginia, is

   a. the oldest continuous community of English origin in America

   b. the annual meeting site of the Daughters of the American Revolution

   c. the site of the shipyard that built the Civil War "cheeseboxes" *Merrimac* and *Monitor*

   d. where the Republican Party was formed

49. The windiest and foggiest site on the west coast of the United States south of the Bering Sea is in California. It is

   a. Point Reyes

   b. Cape Vizcaino

   c. Bodega Head

   d. Point Arena

50. The last spike completing construction of the first transcontinental railroad in the United States was driven at Promontory in the state of

   a. Nevada

   b. Utah

   c. South Dakota

   d. Montana

1. **(c)** Georgia, 58,876 square miles, is the largest state in area east of the Mississippi River, but 20 states west of the Mississippi are larger than Georgia, which was one of the original 13 states.
2. **(d)** Because of sea-floor spreading, the Hawaiian Islands, which stretch 1,500 miles across the Pacific Ocean, continue to inch their way northwestward.
3. **(b)** The Hawaiian Islands were named the Sandwich Islands, for the Earl of Sandwich, by British explorer Captain James Cook. Cook was killed there in 1779.
4. **(b)** Alaska, which was the forty-ninth state admitted to the Union (1959), is the nation's largest state in area, 586,412 square miles. In the 1980 census, it reported the fewest number of residents, 401,851.
5. **(d)** About 4,800 square miles of Minnesota's 84,068 square miles are wet. It is the "land of 11,000 lakes."
6. **(d)** Both Tennessee and Missouri border on 8 other states.
7. **(d)** There are between 80,000 and 200,000 Arabs in the Detroit area, the largest Arab population in the United States.
8. **(d)** The motto of Hawaii is *Ua Mau Ke Ea O Ka Aina I Ka Pono*—"The Life of the Land Is Perpetuated in Righteousness."
9. **(d)** Harney Peak, a 7,242-foot-high mountain in the Black Hills, in southwest South Dakota, is not only the highest point in the state but also the highest point in the United States east of the Rocky Mountains.
10. **(a)** In the Great Smoky Mountains range of the Appalachians along the boundary of Tennessee and North Carolina, in the eastern United States, are the largest tract of hardwood and the largest virgin forest of red spruce in the nation.
11. **(c)** Quincy, Massachusetts, was the birthplace of both John Adams, our second President, and his son John Quincy Adams, our sixth President.

12. **(d)** Among the 17,000 known caves in the United States are Kentucky's Mammoth Cave system, which has 298 miles of mapped passages, and New Mexico's Carlsbad Caverns system, with at least 21 miles of passages. One of the caves in the Carlsbad system is so big that it could accommodate 11 football fields. More than 100 caves are open to the public.

13. **(a)** Crater Lake, 1,932 feet deep in the caldera of Mount Mazama, in the Cascades in southern Oregon, is about 6 miles long and 5 miles wide, and remarkable for the intensity of its blue water.

14. **(c)** One in 9 Americans now live in California, the nation's most populous state (29 million people).

15. **(b)** The 11 states of the corn belt are Illinois, Indiana, Iowa, Kansas, Michigan, Minnesota, Missouri, Nebraska, Ohio, South Dakota, and Wisconsin. Corn reigns over a quarter of all America's cropland; about a third of the product is exported.

16. **(d)** There are, percentagewise, more Hispanic people in New Mexico than in any other state.

17. **(a)** "Sucker State" and "Prairie State" have been nicknames of Illinois, which was admitted to the Union on December 3, 1818, when its capital was Kaskaskia.

18. **(b)** In the 1980 census, blacks made up 11.8 percent of the total population of the United States.

19. **(a)** The "German coast" is a district extending about 40 miles along the right bank of the Mississippi River from a point about 30 miles north of New Orleans.

20. **(c)** President Theodore Roosevelt was active, if not hyperactive, in the early days of the American conservation movement.

21. **(a)** Among the states, South Carolina has the highest infant-mortality rate per 1,000 live births (all races) in the first year of life. (The rate for the District of Columbia is even higher.)

22. **(c)** The 5 states of the United States that border on the Gulf of Mexico are Alabama, Florida, Louisiana, Mississippi, and Texas.

23. **(c)** Minnesota's watersheds are in a real sense "the mother of three seas." Channels run to Hudson Bay and the Arctic, to the Atlantic Ocean, and to the Gulf of Mexico.

24. **(b)** Barrow, Alaska, is the northernmost town in the United States.

25. **(a)** Victory in the French and Indian war assured British political control over their 13 Atlantic seaboard colonies.

26. **(d)** There are 8 main islands in the 132-island chain of the Hawaiian Islands.

27. **(b)** Cincinnati, on the Ohio River in the southwest corner of Ohio, became a grape culture center and a wine market, and was hailed as the "queen of the west" with the opening of the Miami and Erie Canal in 1832.

28. **(d)** Because of increased trade with the Far East, Los Angeles has replaced New York-New Jersey as the nation's busiest port.

29. **(d)** One can drive south from Detroit, Michigan, to enter Canada.

30. **(a)** Badwater, a small pool in Death Valley, in eastern California, is the lowest point in the United States: 282 feet below sea level; Mount Whitney, a peak in the Sierra Nevada, in southeast-central California, is the highest point, 14,494 feet. Badwater and Whitney are only 80 miles apart.

31. **(c)** Colorado's capital city, Denver, is "the mile-high city."

32. **(d)** The state on the Atlantic coast of the United States that is closest to California is Georgia.

33. **(d)** The United States ranks nineteenth lowest in infant mortality (all races) among all the nations of the world with 9.7 deaths per 1,000 live births.

34. **(d)** The 1,038-mile-long Snake River rises in Yellowstone National Park, in northwest Wyoming, and empties into the Columbia River, in southeast Washington. It has carved out a canyon more than 40 miles long and more than 7,000 feet deep at the Idaho-Oregon boundary.

35. **(a)** Rose, an uninhabited island in American Samoa, is 1,000 miles south of the Equator and is the southernmost US territory.

36. **(a)** Hawaii, the southernmost of all states of the United States, is the state with the wettest climate.

37. **(a)** A desertlike region covers much of the American southwest and reaches deep into the Mexican state of Sonora.

38. **(c)** Exactly 2,897 miles separate West Quoddy Head, Maine, and Point Arena, California, the points farthest apart in the 48 contiguous states of the United States.

39. **(a)** The Toledo War of the mid-1830s was a dispute between the states of Ohio and Michigan over the location of their common boundary.

40. **(b)** Immigrants, 7 million to 9 million, both legal and illegal, are driving the current US population increase. They are largely from Latin America, Asia, and the Caribbean. As much as 30 percent of California's growth has been due to immigration.

41. **(b)** Corn and soybeans are the two most important crops grown in Iowa, Illinois, and Indiana.

42. **(d)** Tombstone, in the southeast corner of Arizona, was known as a mining center, as well as for its crime and lawlessness.

43. **(a)** Peshtigo, in northeast Wisconsin, and Chicago, Illinois, were both practically destroyed by fires on October 8, 1871.

44. **(d)** The Appalachian National Scenic Trail in the eastern United States extends from Maine to Georgia.

45. **(b)** The permanent capital of the United States, Washington, D.C., was created from portions of Maryland and Virginia. Its square mileage is 67 and its population is about 650,000. (During the Second World War, its population soared to about 1.25 million. D.C. residents were granted national suffrage through ratification of the Twenty-third Amendment to the Constitution in 1961.)

46. **(a)** The state of Indiana's motto is "The Crossroads of America."

47. **(d)** Devised originally to settle a boundary dispute between the Calvert-family proprietors of Maryland and the Penn-family proprietors of Pennsylvania, the Mason-Dixon line became in 1779 the symbolic border between north and south in the United States, politically and socially, dividing the free states from the slave states.

48. **(a)** The oldest continuous community of English origin in America is Hampton, in southeast Virginia, 7 miles northeast of Newport News.

49. **(a)** Point Reyes, at the southern extremity of a peninsula in Marin County, California, averages 137 days of fog annually and is said to be the windiest and foggiest place on the west coast of the United States south of the Bering Sea.

50. **(b)** To the shout of "DONE!!!," the last spike in construction of the first transcontinental railroad in the United States was driven at Promontory, Utah, about 30 miles west of the city of Brigham, on May 10, 1869. The site is commemorated by the Golden Spike Monument.

MQ24. Name the 10 labeled states of the United States:

1. _____   6. _____
2. _____   7. _____
3. _____   8. _____
4. _____   9. _____
5. _____   10. _____

MQ25. The area within the grey-tone lines on this map of the United States represents

    a. the Northwest Ordinance

    b. land ceded by the Treaty of Guadalupe Hidalgo

    c. the Gadsden Purchase

    d. the Louisiana Purchase

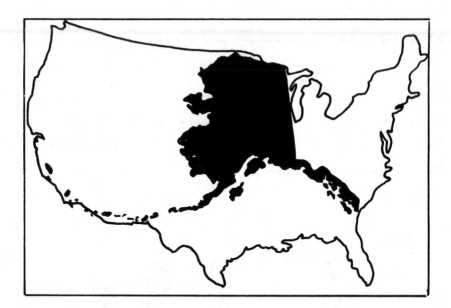

MQ26. A map of the state of _____ lies to scale over a map of the contiguous 48 states of the United States.

    a. Hawaii

    b. Alaska

    c. Texas

    d. Minnesota

MQ27. This hilly-to-mountainous state joined the Union during the Civil War:

    a. Utah

    b. West Virginia

    c. Arkansas

    d. Louisiana

MQ28. This is a map of the third state to join the Union after the
Constitution was ratified:
  a. Tennessee
  b. Kentucky
  c. Ohio
  d. Virginia

MQ29. This is a map of the state of
  a. Nebraska
  b. Oklahoma
  c. Tennessee
  d. Kentucky

MQ30. This semi-tropical state is one of the many along which or
through which the Mississippi River flows:
  a. Arkansas
  b. Mississippi
  c. Louisiana
  d. Illinois

MQ31. This north-central state of 2 peninsulas has a land area of 56,817 square miles and a Great Lakes area of 38,575 square miles:

    a. Michigan

    b. Indiana

    c. Wisconsin

    d. Minnesota

MQ32. This state was a district of another state until 1820:

    a. Massachusetts

    b. Vermont

    c. Maine

    d. Idaho

MQ33. This western state is seventh in area, forty-third in population:

    a. New Mexico

    b. Oregon

    c. Nevada

    d. Montana

MQ34. These 2 adjoining states are:

    a. Wyoming (1) and Colorado (2)

    b. Arizona (1) and New Mexico (2)

    c. Mississippi (1) and Alabama (2)

    d. Wyoming (1) and South Dakota (2)

MQ35. The state is _____ and the white area in the northwest is _____.

    a. Utah . . . the Great Salt Lake

    b. Vermont . . . the Green Mountains

    c. North Dakota . . . the geographical center of North America

    d. New York . . . the Finger Lakes

MQ36. Number 1 on the map pinpoints a city on the west coast of the United States. The city is

    a. San Francisco, California

    b. San Diego, California

    c. Seattle, Washington

    d. Portland, Oregon

MQ24. 1. Mississippi; 2. Idaho; 3. Minnesota; 4. Colorado; 5. South Carolina; 6. Vermont; 7. Oklahoma; 8. Indiana; 9. Nebraska; 10. Arizona.

MQ25. **(d)** For two and a half cents an acre, the United States in 1803 purchased from France the Louisiana Territory: 565,166,080 acres, or 883,072 square miles. It is an area larger than the combined areas of France, Germany, Italy, Spain, and Portugal, and it is 7 times larger than the combined areas of England, Scotland, and Ireland.

MQ26. **(b)** Alaska, the first US state in area and the fiftieth in population, has a land area of 571,065 square miles. The 50 states total 3,615,123 square miles.

MQ27. **(b)** The east-central Mountain State of West Virginia was part of Virginia until the Civil War. It voted against secession in May, 1861, and joined the Union 2 years later. The Allegheny Plateau covers two thirds of the state, whose motto is "Mountaineers Are Always Free Men."

MQ28. **(a)** The southwest-central state of Tennessee, the Volunteer State, joined the Union in 1796. It was the first of the seceding states to be reorganized and readmitted to the Union after the Civil War. Tennessee's climate is humid continental to the north, humid subtropical to the south.

MQ29. **(b)** The west-southcentral state of Oklahoma, the forty-sixth to join the Union (1907), is the eighteenth state in area, at 69,919 square miles. The Sooner State's highest point, Black Mesa, at 4,973 feet, is in the western panhandle. The state's east-central area is dominated by the Arkansas River Basin and the Red River Plains.

MQ30. **(b)** The state is Mississippi, the Magnolia State. The Mississippi, "Ol' Man River," flows along the northern section of the boundary of Louisiana and Mississippi, then continues southeast into the river's delta through several mouths, known as Passes, and into the Gulf of Mexico.

MQ31. **(a)** Michigan, the twenty-sixth state to join the Union, is bounded by several states and the Great Lakes Superior, Huron, Michigan, and Erie and on the east by Canada. The Upper and Lower Peninsulas are separated by the Straits of Mackinac, which links Lakes Michigan and Huron. Michigan's motto is "If You Seek a Beautiful Peninsula, Look Around You."

MQ32. **(c)** The northeast Pine Tree State of Maine was annexed to Massachusetts in 1652. It was a district of the Bay State until 1820, when it was admitted to the Union as a non-slave state. The harsh northern region averages more than 100 inches of snow every winter.

MQ33. **(c)** Semi-arid Nevada, at 110,540 square miles, is the seventh state in area, the forty-third in population. Its highest point is Boundary Peak, 13,140 feet, on the Nevada-California boundary. The Sagebrush State's southern area is within the Mojave Desert.

MQ34. **(a)** The states are Wyoming (1), "the Equality State," and Colorado (2), "the Centennial State." Together, these 2 western areas have an aggregate of 32,299,300 acres of forested land.

MQ35. **(a)** The Rocky Mountain state is arid Utah, "the Beehive State," and the white area in the northwest is the Great Salt Lake, which was discovered in 1824. The region was acquired by the United States from Mexico in the Treaty of Guadalupe Hidalgo in 1848, and Utah joined the Union 48 years later. The southeastern corner of Utah touches Arizona, New Mexico, and Colorado; it is the only place in the nation where 4 states adjoin.

MQ36. **(a)** San Francisco (46 square miles), on the west side of San Francisco Bay, is a California seaport city with an exceptionally good harbor.

- The archipelago Comoros, off Africa's eastern coast, is the world's leading producer of the essence of ylang-ylang, used in manufacturing perfume. The literacy rate there is 15 percent, and life expectancy is nearly 49 years.
- The Soviet Union has a greater presence in Peru than in any other South American country.
- Life expectancy for women in Japan is 81.4 years; for men, 75.6.
- Lake Villafro, a lake in southern Peru, is regarded as the remotest source of the Amazon River.
- In the 250 years from 1565 to 1815, Acapulco was Spain's western-hemisphere seaport for trade with the Philippines. "Manila galleons" made yearly voyages across the Pacific; goods were transshipped in Mexico.
- Scandinavia's high mountain range keeps Atlantic air from sweeping into Sweden and also directs Arctic conditions southward into the heart of the Scandinavian peninsula.
- There are millions of North Africans living in France, Turks in Germany, Surinamese in the Netherlands, and Indians and Pakistanis in Great Britain—about 10 percent of western Europe's population is now an immigrant population, mostly from Third World countries.
- Charybdis, the famous Mediterranean whirlpool near Cape Faro, Sicily, is now called Galofalgo.
- Athabasca, a river in west-central Canada, has one of the largest oil reservoirs in the world.

1. In ancient geography, the Erythraean Sea was part of the
   a. Atlantic Ocean
   b. Mediterranean Sea
   c. Dead Sea
   d. Indian Ocean

2. The largest freshwater lake in the world is
   a. Superior
   b. Michigan
   c. Victoria
   d. Bato

3. Europe's longest river is the
   a. Volga
   b. Danube
   c. Po
   d. Thames

4. The Pacific is the world's largest ocean. The second largest is the
   a. Atlantic
   b. Indian
   c. Arctic
   d. Southern

5. The only river in the Arab Near East that does not cross a national boundary is the
   a. Litani
   b. Euphrates
   c. Jordan
   d. Tigris

6. The Magdalena, the São Francisco, and the Orinoco are major
_____ in South America.

    a. rivers

    b. seas

    c. lakes

    d. reservoirs

7. Ishikari is

    a. the name of the underwater city under construction in Tokyo Bay

    b. the name of North Korea's College of Oceanography

    c. the group name of 5 undersea volcanic mountains in the northern Philippines

    d. the second longest river in Japan

8. Rio Roosevelt, a 400-mile-long river in west-central Brazil, which was explored by former President Theodore Roosevelt in 1914, was once known as

    a. the River of Gold

    b. the River of Doubt

    c. the River of Disappointment

    d. the River of Great Happiness

9. In Numbers in the Old Testament, "the Great Sea" refers to

    a. the Dead Sea

    b. the Nile River

    c. the Indian Ocean

    d. the Mediterranean Sea

10. The headwaters of the longest river in the United States are in the state of

    a. Minnesota   **c.** Ohio

    b. Alaska   **d.** Oregon

11. The cardinal direction that most of the major rivers in the Great Plains of the United States flow is
    a. north
    b. east
    c. south
    d. west

12. Only 1 of the 5 Great Lakes is located entirely within the United States:
    a. Superior
    b. Michigan
    c. Huron
    d. Erie

13. The Volga and the Danube are the 2 longest rivers in Europe. The continent's third longest river is the
    a. Dnieper
    b. Rhone
    c. Seine
    d. Po

14. "The graveyard of the Atlantic" is the sea off this cape:
    a. Cod
    b. Hatteras
    c. Good Hope
    d. Horn

15. The only Canadian province that borders the Great Lakes is
    a. Quebec
    b. Ontario
    c. Alberta
    d. Saskatchewan

16. Norway's longest river, Glama, is _____ miles long.
    a. 33
    b. 106
    c. 380
    d. 993

17. Four mighty rivers—the Niger, the Nile, the Congo, and the Zambezi—rise in the high interior of
    a. Brazil
    b. Thailand
    c. the Soviet Union
    d. Africa

18. The world's longest river system is
    a. the Nile
    b. the Danube
    c. the Volga
    d. the Amazon

19. The deepest freshwater lake in the world is
    a. Baikal
    b. Titicaca
    c. Victoria
    d. Geneva

20. The waterway Shatt-al-Arab, which is formed by the confluence of the Tigris and Euphrates Rivers, has been the subject of a long-standing border dispute between Iraq and
    a. Iran
    b. Kuwait
    c. Afghanistan
    d. the Soviet Union

21. These 4 seas—the Adriatic, the Ionian, the Ligurian, and the Tyrrhenian—all wash ashore onto
    a. Italy
    b. Christmas Island
    c. Taiwan
    d. Saudi Arabia

22. Forty-six percent of the Earth's water is in
    a. glaciers
    b. the Pacific Ocean
    c. the waters north of the Arctic Circle and south of the Antarctic Circle
    d. lakes and rivers

23. Tides are caused by
    a. the planets' orbital positions
    b. gravity, by the attraction of the Moon and Sun
    c. the local magnetic fields
    d. the movement of whales

24. The largest natural lake in Africa is
    a. Lake Victoria
    b. Lake Tanganyika
    c. Lake Malawi
    d. the Aral Sea

25. The Nile is the longest river system in the world: 4,132 miles. The Amazon is the second longest: 4,000 miles. At 3,434 miles, the _____ is the world's third longest.
    a. Yangtze
    b. Mississippi
    c. Danube
    d. Thames

26. The world's highest tides are in the Bay of Fundy, an inlet of the Atlantic Ocean in
   a. Antarctica
   b. Canada
   c. Greenland
   d. Iceland

27. The delta of the Rhone River is in
   a. Belgium
   b. the Netherlands
   c. France
   d. Poland

28. The _____ River touches more countries than does any other river in the world.
   a. Danube
   b. Nile
   c. Rhine
   d. Seine

29. About _____ of water from the Amazon River discharges into the Atlantic Ocean every second.
   a. 100 cubic feet
   b. 1,000 cubic feet
   c. a half million cubic feet
   d. 7 million cubic feet

30. The Gulf of Guinea borders on
   a. western Africa
   b. western Australia
   c. northeastern Brazil
   d. western France

31. Three of the Soviet Union's main rivers—the Lena, the Ob, and the Yenisey—are not considered practical for transportation, because they all
   a. flow north to the Arctic
   b. flow east into Siberia
   c. tend to be blocked with silt at least twice a year
   d. are much too shallow in long stretches

32. The Mississippi River touches _____ states.
   a. 5
   b. 9
   c. 10
   d. 13

33. The smallest ocean is the
   a. Arctic
   b. Indian
   c. Atlantic
   d. Pacific

34. The Bering Sea, named for a Danish explorer, is part of the
   a. Baltic Sea
   b. North Sea
   c. Atlantic Ocean
   d. Pacific Ocean

35. Solo is the name of the
   a. longest river in Java
   b. sea in Wales
   c. only lake in Somalia
   d. current surrounding Australia

36. The White Nile and the Blue Nile come together at the city of
    a. Cairo
    b. Khartoum
    c. Luxor
    d. Malakal

37. The most important tributary of the Amazon River is _____ miles long.
    a. 530
    b. 1,140
    c. 2,013
    d. 3,172

38. The Mississippi is the longest river in North America. The second longest is the
    a. Missouri
    b. Fraser
    c. Mackenzie
    d. Snake

39. The source of the Mississippi River is Lake
    a. Superior
    b. Michigan
    c. Itasca
    d. Leech

40. The water level in the Soviet Union's Aral Sea is rapidly dropping because of
    a. evaporation
    b. the withdrawal of water for farming
    c. frequent earthquakes in the region
    d. a decade-long drought

41. The most vital commercial waterway in Europe is the
    a. Seine
    b. Elbe
    c. Danube
    d. Rhine

42. The Mississippi River is to New Orleans as the Rhine River is to
    a. Rotterdam
    b. Kiel
    c. Marseille
    d. Budapest

43. The only major ice-free port in the Soviet Union that is not blocked from the open ocean by narrow straits is on
    a. the Bering Sea
    b. the Arctic Ocean
    c. the Pacific Ocean
    d. the Indian Ocean

44. The longest artificially created waterway in the world is
    a. the Grand Canal, in the People's Republic of China
    b. the Nord-Ostsee Kanal (the Kiel Canal), in Germany
    c. the All-American Canal, in the state of California
    d. the Canal du Midi, in France

45. Hudson Bay is an inland sea in
    a. the Northwest Territories, Canada
    b. New York State
    c. the Netherlands
    d. Antarctica

46. The largest lake in England, Windermere, in the Lake District in the northwest part of the country, is _____ miles long.

    a. 1.5

    b. 33

    c. 49

    d. exactly 99.5

1. **(d)** The Erythraean Sea, in ancient geography, was part of the Indian Ocean known as the Persian Gulf and Arabian Sea.

2. **(a)** One of the Great Lakes, Superior, is the largest body of fresh water in the world: 31,800 square miles. The US-Canadian boundary passes through the middle of the lake.

3. **(a)** The Volga, Europe's longest river, flows 2,290 miles across the Soviet Union and empties into the Caspian Sea. The continent's second longest river, the Danube, flows 1,776 miles from its source in west Germany to its mouth in the Black Sea.

4. **(a)** The Pacific Ocean is about 70 million square miles. The second largest ocean is the Atlantic, which covers about 32 million square miles. The third largest is the Indian Ocean, which covers about 28 million square miles. The fourth largest is the Arctic Ocean, which covers about 5.5 million square miles. The Southern Ocean, which circles Antarctica, is not officially an ocean.

5. **(a)** The Litani River is wholly within Lebanon. It rises near Baalbeck in southern Lebanon and empties into the Mediterranean Sea 6 miles north of Tyre.

6. **(a)** The Magdalena in Colombia, the São Francisco in east Brazil, and the Orinoco in Venezuela are major South American rivers; they just happen to be dwarfed by the Amazon.

7. **(d)** Ishikari, Japan's second longest river, flows 275 miles in western Hokkaido Island and empties into Ishikari Bay.

8. **(b)** The River of Doubt is the erstwhile name of west-central Brazil's Rio Roosevelt.

9. **(d)** In Numbers in the Old Testament, "the Great Sea" is the Mediterranean Sea: "And as for the western border, ye shall even have the great sea for a border, this shall be your western border."

10. **(a)** The headwaters of the Mississippi, the longest river in the United States, are in the state of Minnesota.

11. **(b)** Most of the major rivers in the Great Plains of the United States flow east.

12. **(b)** Lake Michigan is the only one of the 5 Great Lakes that lies wholly within the United States. The US-Canadian boundary runs through the 4 other Great Lakes: Ontario, Erie, Huron, and Superior.

13. **(a)** The Dnieper, 1,420 miles long, is Europe's third longest river and is second only to the Volga, Europe's longest river, as the longest river in the Soviet Union. The Dnieper rises near the source of the Volga, and, like the Danube, Europe's second longest river, it empties into the Black Sea. In 1667, the Dnieper became the boundary between Russia and Poland.

14. **(b)** The sea off the long, narrow sandbar that is Cape Hatteras, North Carolina, is known as "the graveyard of the Atlantic," because so many ships have been wrecked there; it is a dangerous area for navigation.

15. **(b)** Ontario is the only Canadian province that borders on the Great Lakes.

16. **(c)** Norway's longest river, the Glama, is 380 miles long, flowing through the eastern part of the country.

17. **(d)** The Niger, Nile, Congo, and Zambezi Rivers rise in the high interior of Africa. Each has impassable rapids, making boat traffic from one end to the other difficult.

18. **(a)** Africa's Nile is the world's longest river system: 4,132 miles.

19. **(a)** The Soviet Union's Baikal, with its maximum depth of 5,715 feet, is the deepest freshwater lake in the world.

20. **(a)** Iran and Iraq have gone to war over the waterway known as Shatt-al-Arab.

21. **(a)** The boot-shaped Italian peninsula in southern Europe is washed by the Adriatic, Ionian, Ligurian, and Tyrrhenian Seas.

22. **(b)** The Pacific Ocean—about 70 million square miles—contains 46 percent of the Earth's water.

23. **(b)** Gravity—stimulated by the attraction of the Moon and Sun—is the major force that creates tides.

24. **(a)** Lake Victoria, 26,828 square miles, is the largest natural lake in Africa. Lake Tanganyika is the second largest: 12,703 square miles. Lake Malawi is the third largest: 11,158 square miles. (The Aral Sea is in the Soviet Union.)

25. **(a)** The Yangtze, or Chang Jiang, the third longest river in the world, flows 3,434 miles through the People's Republic of China.

26. **(b)** The Bay of Fundy is an inlet of the Atlantic Ocean between the Canadian provinces of Nova Scotia and New Brunswick, and is the site of the world's highest tides.

27. **(c)** The delta of the Rhone River is in the French region known as the Camargue.

28. **(a)** The fabled Danube River originates in southern Germany, drifts lazily in a southeastern direction, and empties into the Black Sea, touching 8 countries along the way. It is continually fed by befouled tributaries, and in turn is a major polluter of the Black Sea.

29. **(d)** The outpouring of the Amazon River, whose source is in the Peruvian Andes, is so great—7 million cubic feet every second—that the open sea is freshwater for over 200 miles beyond the mouth of the Amazon.

30. **(a)** The Gulf of Guinea borders on western Africa.

31. **(a)** Because they all flow north to the Arctic, away from resources and population, the Soviet Union's Lena, Ob, and Yenisey Rivers are not considered practical for transportation.

32. **(c)** The Mississippi River touches 10 states: Arkansas, Illinois, Iowa, Kentucky, Louisiana, Minnesota, Mississippi, Missouri, Tennessee, and Wisconsin.

33. **(a)** The Arctic is the smallest ocean, "only" 3,662,177 square miles. The second smallest is the Indian: 28,350,311 square miles.

34. **(d)** The Bering Sea—885,000 square miles, with a maximum depth of 15,659 feet—is part of the North Pacific. It is crossed by the International Date Line.

35. **(a)** The Solo River in Java is 335 miles long, the Indonesian island's longest river.

36. **(b)** It is at the city of Khartoum, the Sudan, that the White Nile and the Blue Nile come together and form the Nile. The Blue (about 1,000 miles) originates in northern Ethiopia, the White (about 1,650 miles) in Lake Victoria.

37. **(c)** The Madeira, the most important tributary of the Amazon River, is 2,013 miles long (when combined with the Mamore). It rises in western Brazil and flows northeast into the Amazon below Manaus.

38. **(c)** The second longest river in North America, the Mackenzie in the Northwest Territories, Canada, is about 2,635 miles long when the Slave, Peace, and Finlay Rivers are included.

39. **(c)** Lake Itasca, 2 square miles in northern Minnesota, is generally referred to as the source of the Mississippi River. Its name is from two Latin words: *veritas* (truth) and *caput* (head).

40. **(b)** A human activity—irrigation, the withdrawal of water for farming—is responsible for the rapid drop in the water level of the Aral Sea in the Soviet Union.

41. **(d)** The Rhine, 820 miles long, is considered the most vital commercial waterway in Europe.

42. **(a)** New Orleans is at the mouth of the Mississippi River, and Rotterdam, the Netherlands, is at the mouth of the Rhine River.

43. **(b)** The Soviet Union's Arctic port of Murmansk remains ice-free because of the North Atlantic Drift and the Gulf Stream. It is the country's only major ice-free port that is not blocked from the open ocean by narrow straits.

44. **(a)** Construction of the longest artificially created waterway in the world—about 1,000 miles—China's Grand Canal—was begun more than 2,400 years ago. It is a system of canals and navigable sections of major rivers, including the Yangtze and the Yellow.

45. **(a)** Hudson Bay, whose maximum depth is 2,846 feet, is an inland sea in the eastern section of Canada's Northwest Territories.

46. **(a)** Windermere, the largest lake in England, is 1.5 miles long, with a maximum depth of 219 feet.

Major Earthquakes Since 1966

Geography is often at the mercy of Mother Nature:

Year	Place	Deaths
1990	Philippines	1,200
1990	Iran	40,000
1989	California	62
1988	Armenia	25,000
1985	Mexico	9,500
1983	Turkey	1,300
1982	Yemen	2,800
1980	Italy	4,800
1980	Algeria	4,500
1979	Colombia, Ecuador	800
1978	Iran	25,000
1977	Romania	1,541
1976	Turkey	4,000
1976	Philippines	8,000
1976	China	upward of 800,000
1976	Italy	946
1976	Guatemala	22,778
1975	Turkey	2,312
1974	Pakistan	5,200
1972	Nicaragua	5,000
1972	Iran	5,057
1970	Peru	66,794
1970	Turkey	1,086
1968	Iran	12,000
1966	Turkey	2,520

In 1927 an earthquake killed 2,000,000 people in China; in 1923 a quake killed 100,000 in Japan; and in 1920 a quake killed 100,000 in China. In 1908, 83,000 people died in a quake in Italy.

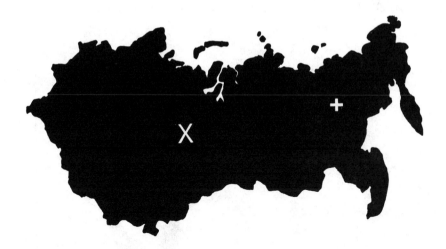

MQ37. The Soviet Union's vast, far-eastern, frozen Autonomous Republic of Yakutia+, where winter temperatures plunge to minus 50 degrees Fahrenheit, vies with southern Africa's Botswana as the world's leading producer of

    a. oil barrels

    b. diamonds

    c. uranium

    d. winter wheat

MQ38. X, which is about 900 miles east of Moscow, marks

    a. the Soviet Union's ICBM launching pads

    b. the most polluted spot on the planet

    c. a secret training center for Olympic competitors

    d. the documented site of an extraterrestrial visitation

MQ39. The diagonally marked area in this map of Israel represents
   a. the West Bank
   b. the Dead Sea
   c. the Gaza Strip
   d. the Golan Heights

MQ40. The United States has about 90 phone lines per 100 people. Western Europe has about 60. Eastern Europe averages only 9 phones for every 100 people. Place an X on the European country where a citizen must wait 13 years to get a telephone in his home.

MQ41. The 5 horizontal lines from north to south represent the
Arctic Circle (1), the Tropic of Cancer (2), the Equator (3),
the Tropic of Capricorn (4), and the Antarctic Circle (5).
On what lines are these 8 countries:

Mozambique _____

the Soviet Union ___

Colombia _____

Paraguay _____

Finland _____

Mexico _____

Saudi Arabia _____

Brazil _____

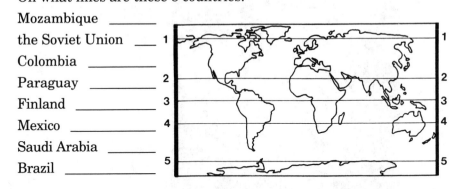

MQ42. The 10 countries with the largest crude oil reserves in
mid-1990 were Venezuela, United Arab Emirates, Saudi
Arabia, the People's Republic of China, the United States,
the Soviet Union, Iraq, Iran, Kuwait, and Mexico. Rank
them on the map from 1, the largest, to 10.

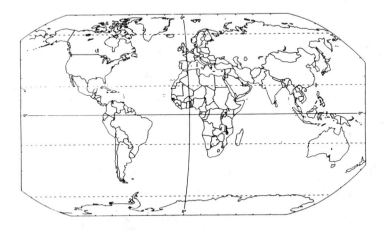

MQ43. These 7 countries consumed the most oil in mid-1990:
Japan, France, the United States, Britain, Italy, Germany,
Canada. Rank them on the map from 1, the largest
consumer, to 7.

MQ44. Match the 7 labeled places in North America:

Great Slave Lake	1. _____
Mississippi River	2. _____
Great Basin	3. _____
Yukon	4. _____
Rocky Mountains	5. _____
Appalachians	6. _____
Western Sierra Madre	7. _____

MQ45. Name the 5 Great Lakes in central North America:

1. _____
2. _____
3. _____
4. _____
5. _____

MQ46. If a map of Europe were placed to scale over a map of the United States, match the 13 numbers with the location of these major European cities:

Budapest	1. _____
Prague	2. _____
Warsaw	3. _____
Berlin	4. _____
Rome	5. _____
Moscow	6. _____
London	7. _____
Madrid	8. _____
Athens	9. _____
Lisbon	10. _____
Paris	11. _____
Belgrade	12. _____
Dublin	13. _____

MQ47. The Atlantic Ocean (1) and the Pacific Ocean (2) together are
_____ million square miles.

    a. 7

    b. 16

    c. 73

    d. 101

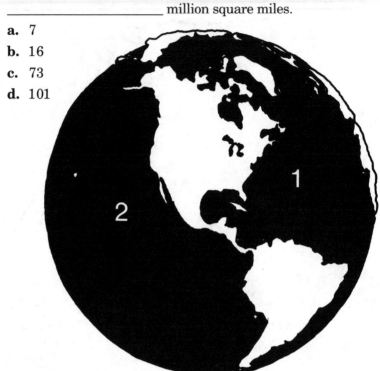

MQ48. This is a map of the most populous country in the world:

    a. the Soviet Union

    b. the People's Republic of China

    c. India

    d. the United States

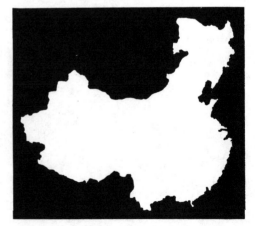

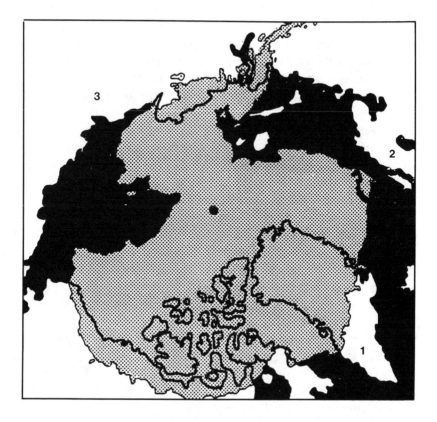

MQ49. The black dot in the center of this layout of superimposed geofeatures represents _____ . The 3 labeled areas are (1) _____ ; (2) _____ ; and (3) _____ .

MQ50. The grey-tone area in the cartography above represents the continent of _____ .

MQ37. **(b)** Yakutia, which is the size of India and home to 1 million hearty people, produces all of the Soviet Union's diamonds (an increasingly valuable source of hard currency) and competes with Botswana as the planet's leading producer of the colorless, hardest natural substance. The northland republic south of the East Siberian Sea in fact produces three quarters of all the USSR's fuel and minerals; the major diamond discoveries were made there in the 1950s.

MQ38. **(b)** The Soviet Union's nuclear weapons production complex at Chelyabinsk (X on the map) was described in 1990 as the most polluted spot on the planet. It has a history of extremely intense radioactive contamination and of very high doses to production workers.

MQ39. **(a)** The diagonally marked West Bank, which is west of the Jordan River, has been an integral part of Israel since the Six-Day War of June, 1967. West Bank cities include Hebron and the Israeli capital, Jerusalem.

MQ40. It takes 13 years for a Polish citizen to get a telephone in his home.

MQ41. Mozambique is on the Tropic of Capricorn; the Soviet Union, the Arctic Circle; Colombia, the Equator; Paraguay, the Tropic of Capricorn; Finland, the Arctic Circle; Mexico, the Tropic of Cancer; Saudi Arabia, the Tropic of Cancer; Brazil, both the Equator and the Tropic of Capricorn.

MQ42. The countries in mid-1990 with the largest crude oil reserves (in billions of barrels) were: 1. Saudi Arabia, 255.0; 2. Iraq, 100.0; 3. United Arab Emirates, 98.1; 4. Kuwait, 94.5; 5. Iran, 92.9; 6. Venezuela, 58.5; 7. the Soviet Union, 58.4; 8. Mexico, 56.4; 9. the United States, 25.9; 10. the People's Republic of China, 24.0.

MQ43. The 7 countries in mid-1990 with the biggest appetites for oil were: 1. the United States; 2. Japan; 3. Germany; 4. Italy; 5. France; 6. Canada; 7. Britain.

MQ44. The 7 labeled North American places are: 1. Appalachians; 2. Rocky Mountains; 3. Western Sierra Madre; 4. Mississippi River; 5. Great Basin; 6. Yukon; 7. Great Slave Lake.

MQ45. The Great Lakes are: 1. Lake Superior, about 350 miles long, the largest body of fresh water in the world, 31,800 square miles; 2. Lake Huron, the second in size of the Great Lakes, about 260 miles long, 23,000 square miles; 3. Lake Erie, about 241 miles long, 9,910 square miles, the fourth in size, its greatest depth 210 feet; 4. Lake Michigan, about 307 miles long, 22,400 square miles, third in size and the only Great Lake wholly within the United States; 5. Lake Ontario, the easternmost and the smallest in size, 193 miles long, 7,600 square miles.

MQ46. The 13 European cities are: 1. Lisbon; 2. Madrid; 3. Dublin; 4. London; 5. Paris; 6. Rome; 7. Berlin; 8. Prague; 9. Warsaw; 10. Budapest; 11. Belgrade; 12. Moscow; 13. Athens.

MQ47. **(d)** The Atlantic and Pacific Oceans together are 101 million square miles. The Pacific is more than twice as big as the Atlantic.

MQ48. **(b)** The People's Republic of China, with more than a billion people, is the most populous country in the world.

MQ49. The black spot represents the North and South Poles overlaid. The labeled areas are: 1. Greenland; 2. Scandinavia; 3. USSR.

MQ50. The grey-tone area in the cartography represents Antarctica.

AN INVITATION

We are planning a sequel to *Where On Earth?* If you have favorite, unexpected, dramatic geographic facts that aren't in this book and you'd like to share them, we would be pleased to have them in hand when we prepare our next book. Send such facts, with a documented source, to me in care of Prentice Hall Press, 15 Columbus Circle, New York, New York 10023. Enjoy this book, and send along your special goodies for the next one. Thanks. Donnat V. Grillet.

This Index lists the principal references in the 14 all-text multiple-choice sections and in Oddish Facts.

ABOUT THE AUTHOR

A master's graduate of Temple University, Donnat V. Grillet has been since 1965 a teacher and a curriculum collaborator in the Philadelphia school district, developing geography curricula and teaching methods for teachers and schools. He has published articles on geography in many periodicals, including the *Philadelphia Inquirer*'s Newspaper in Education. His geographic puzzles have appeared in the *Philadelphia Daily News*. The author lives with his wife, Grace, and children, Beth and Don, in Havertown, Pennsylvania. He says, "We're making geography fun here in Philadelphia."